Astronomical Alchemy

How the Universe Built Itself and Us

Geoff Canham, PhD

Astronomical Alchemy

CONTENTS

Contents

Astronomical Alchemy

x

Image Credits

Almost all the images are courtesy of NASA, who make them freely available for most uses. But don't blame NASA for any of the text within the book. You can see the full color version of the images, along with more information about them, by searching the NASA website at https://nasa-search.nasa.gov/search/images?affiliate=nasa&query=

Introduction: Hubble Space Telescope being released from Space Shuttle Discovery in 1990. Image credit: NASA

Chapter 1: A population of growing black holes over 13 billion light-years from Earth. Image credit: X-ray: NASA/CXC/Univ. of Rome/E.Pezzulli et al. Illustration: NASA/CXC/M.Weiss

Chapter 2: The Necklace Nebula 15,000 light-years away in the constellation Sagitta. Image credit: NASA, ESA, and the Hubble Heritage Team (STScI/AURA)

Chapter 3: Composite image of IC 10 with X-ray data from Chandra and an optical image taken by amateur astronomer Bill Snyder from the Heavens Mirror Observatory in Sierra Nevada, California. Image credit: X-ray: NASA/CXC/UMass Lowell/S. Laycock et al.; Optical: Bill Snyder Astrophotography

Chapter 4: Dark matter in the center of the giant galaxy cluster Abell 1689 that contains about 1,000 galaxies and trillions of stars. Image credit: NASA, ESA, D. Coe (NASA Jet Propulsion Laboratory/California Institute of Technology, and Space Telescope Science Institute), N. Benitez (Institute of Astrophysics of Andalusia, Spain), T. Broadhurst (University of the Basque Country, Spain), and H. Ford (Johns Hopkins University)

Chapter 5: Cosmic Microwave Background (CMB). Image credit: ESA/Hubble & NASA, T. Kitayama (Toho University, Japan)/ESA/Hubble & NASA.

Chapter 6: M51, a spiral galaxy about 30 million light-years away merging with a smaller galaxy. Image credit: X-ray: NASA/CXC/SAO; Optical: Detlef Hartmann; Infrared: NASA/JPL-Caltech

Chapter 7: Star-forming region NGC 2174. Image Credit: NASA/JPL-Caltech

Chapter 8: Spiral galaxy M81 with a supermassive black hole at its center that is about 70 million times as massive as our Sun. Image Credit: X-ray: NASA/CXC/Wisconsin/D.Pooley & CfA/A.Zezas; Optical: NASA/ESA/CfA/A.Zezas; UV: NASA/JPL-Caltech/CfA/J.Huchra et al.; IR: NASA/JPL-Caltech/CfA

Chapter 9: Saturn as seen from Voyager 1 on Nov. 16, 1980. Image Credt: NASA/JPL

Chapter 10: A full Moon above the Earth's horizon as seen from the International Space Station. Image Credit: NASA

Chapter 11: Artist's conception of a catastrophic collision between two rocky exoplanets. Image Credit: NASA/SOFIA/Lynette Cook

Chapter 12: Heavy loads of sediment, nutrients, and dissolved organic material flowing into the Atlantic Ocean after storms in South Carolina. Image Credit: NASA/Goddard/SuomiNPP/VIIRS via NASA's OceanColor

Chapter 13: The Oakmont Stegosaurus. Image credit: The author of this book — nobody else is to blame.

Chapter 14: The Dragonfly mission, due to launch in 2026, will fly multiple sorties to sample and examine sites around Saturn's icy moon Titan, looking for prebiotic chemical processes common on both Titan and Earth. Image Credit: NASA/JHU-APL

Chapter 15: Supernova 1987A with its debris beginning to impact the surrounding ring. Image Credit: NASA, ESA, and P. Challis (Harvard-Smithsonian Center for Astrophysics)

Chapter 16: Planck's view of the whole sky. Image Credit: ESA, HFI & LFI consortia (2010)

Chapter 17: The galaxy NGC 4388 being transformed by interactions with other galaxies in the Virgo Cluster. Image credits: ESA/NASA

Appendix A: Images by author, as you can probably tell

Astronomical Alchemy

xiv

Introduction

In this book, I will be looking at how the universe created gold (and all the other elements) and created life. Interestingly, we'll see that the universe followed the same general pattern as the ancient alchemists did, but then many of the ancient astronomers were also alchemists, so perhaps that shouldn't be surprising. In a lot of ways, we can say that the universe is the original alchemist, creating all the elements including gold, and then creating lifeforms including us. The big question is: What does the universe have up its sleeve for its next trick? Don't expect definitive answers to that question here, but I may indulge in some baseless speculation (or B.S. for short).

One word of warning: Although I have been fascinated by and have studied astronomy and cosmology for more decades than I wish to declare, I am in no way a professional astronomer or cosmologist. When I say things like "we have discovered" in this book, I am using "we" in the sense of "humanity" and not including myself in the subgroup of humanity that carried out the activities leading to those discoveries. And please look at any of my personal thoughts and suggestions as being offered mainly as ideas that may inspire better ideas or might be better simply ignored.

I should also emphasize the difference between knowledge and opinion and point out that there is a spectrum between these two. True science never really accepts anything as fully known beyond dispute but will always be testing to see if an idea really is true. Things that have been "proven" to be "true" in the past have later been found to be not the whole truth or even be plain wrong. On the other hand, some people firmly believe that their opinions on a subject are beyond dispute, even if they have no backup evidence to substantiate it. Our understanding of most of the cosmos and its history falls somewhere along the spectrum between knowledge and opinion because (1) most of the universe is way too far distant to carry out detailed studies and experiments, and (2) a lot of the events happened in the dim and distant past and we don't have time machines to go back and record what happened. But we can try applying the knowledge that we gained from studying nearby objects and devise tools, like the James Webb Space Telescope, to gain as much data on distant objects as we can. We can also use computer modeling to test ideas and see how well the results compare with what we observe. Nevertheless, it is virtually certain (or is it just my opinion?) that every time we send a probe to a distant region of the cosmos for the first time, we will continue to be amazed and astounded by the results that get sent back.

Astronomy is historically connected with astrology, and indeed the two terms used to be synonymous, and many of the famous astronomers of the past made their living by casting horoscopes. What spurred the cataloging of stars and the observation of the motion of planets was the desire to produce more accurate astrological charts. Everything in the universe is connected by gravitational, electromagnetic, and other forces that span the universe, so the same forces that drive the planets also drive our lives. The skill of astrology is supposedly involved in interpreting what these kinds of forces are doing in our lives. That idea might be questionable, but these forces can also lead to predictions about how the universe itself is evolving and help predict its future, even if those forces may have dubious implications for what sort of day you are going to have. I do mention some aspects of astrology in this book, if only because anyone interested in astronomy is almost certain to be asked about astrology from time to time, and it's good to have some knowledge of a subject, especially if you are then going to go on and dismiss it.

Tying predictions to the movement of heavenly bodies and the forces driving those movements may have made astrological readings sound like a valid practice. At least two interpretations of quantum theory suggest that everything we know as matter, including us, emerges from a universe-spanning quantum wave, where every point in the wave affects everything else. So, astrologers might claim support from that. However, demonstrating a link between the position of the planets and how good a day you are going to have is certainly not conclusive, by a long shot. Take any reference to astrology and alchemy in this book as being rather tongue-in-cheek, even if people like Sir Isaac Newton were deep into them. If you think about how Homo sapiens have only been around for about 200,000 years while our Sun takes 230 million years to orbit the Milky Way once, if there are astrological patterns to find, we really haven't been around long enough to notice. But some of these kinds of fanciful ideas make astronomy even more interesting, so I'll be giving them passing reference.

Astronomy is about stars, planets, and other wonders beyond Earth, and the Universe is about 13.8 billion years old[i]. That can be hard to visualize, even though when I use the term "billion" I only mean a thousand million, not the older UK "billion", which was a million million. I might also write numbers such as 13.8 billion in the format 1.38 x 10^{10}, which means 1.38 multiplied by ten ten-times-over, or 1.38 multiplied by a number that is a 1 followed by ten zeros. This "scientific notation" uses superscripted numbers which should not be confused with the superscripted Roman numerals used in the text to link with the source references at the end of the book.

To simplify the timescale of the universe, let's follow Neil deGrasse Tyson's example and imagine that the universe is only one Earth-year old. In this "cosmic year", the Big Bang would have occurred as January 1 started, and the universe began smaller than an atom (we'll question that later, but let's accept the idea for now). By January 4, the universe had expanded and cooled enough to allow the first stars and small protogalaxies to form; by January 7 we had recognizable small galaxies and by the 8th of March our Milky Way galaxy was starting to grow. Supernovae occurred as the early large stars died, and that made and spread new atoms and elements throughout the universe. By May 7, the Milky Way had spun itself out into a disk shape. It is not until around September 2 that our Sun is born. Then the gas and dust around the Sun formed the solar system, including Earth. Life began on Earth around September 19, and by October 17, sea life was flourishing. Around October 29, there was the first great dying as oxygen levels built up, poisoning most of the early lifeforms. By November 22, life was dividing itself up into what we now know as plants, fungi, and animals. Development speeds up as animals with backbones emerge on December 16, plants move onto the land on December 17 and animals follow on the 18th. On December 25 there was the worst extinction ever (the Permian-Triassic extinction) which led to dinosaurs coming to dominate the animal kingdom, and they survived until December 30 when the asteroid hit. Humans, in the form of Homo erectus, arrived on the scene at 10:46 pm on December 31 and Homo sapiens (us) made an appearance around 13 minutes before midnight. By two minutes before midnight, we were the only human species remaining after the Neanderthals and Denisovans had died out.

Astronomy has the advantage of being a time-machine that lets us look back in time. We are seeing the stars and galaxies as they were years ago or, quite likely, thousands, millions or billions of years ago, since the light from them has taken that long to reach us because they are so far away. A telescope is the thing that lets us access this time machine.

For instance, looking at the Andromeda galaxy, which is about 2.5 million light-years away, means that you are seeing it as it was when our human ancestors first started using stone tools. One light-year is about 5.88 trillion (5.88 x 10^{12}) miles or 9.46 trillion (9.46 x 10^{12}) kilometers. Looking at the giant star Betelgeuse, which is 640 light-years away, means that you are seeing it around the time when the Peasants' Revolt in England nearly resulted in the abolition of serfdom there. With special telescopes that can observe things in the microwave band of the electromagnetic spectrum, rather than in normal light, we can look back almost to the time of the Big Bang. The James Webb Space Telescope is just starting its work as I write this and, fingers crossed, we'll be getting a lot of spectacular new information from the far distant past.

Famous Astronomers and Their Discoveries

The night sky seems to have fascinated humans from the time we first stood erect on the African savannahs, and the following list has some of the people who have contributed to our knowledge of the cosmos, but the list is certainly not comprehensive. Also, most of their discoveries are the culmination of a line of discoveries and insights by a series of people before them.

Aristotle (384 – 322 BCE): Greek philosopher who saw heavenly bodies moving overhead at different rates and concluded that they were fixed to 53 concentric, transparent crystalline spheres that rotated on various axes around the Earth.

Aristarchus of Samos (310 – 230 BCE): Probably the first person to consider that the Earth might revolve around the Sun, rather than vice versa, and he also stated that stars were similar objects to the Sun, just a lot further away.

Eratosthenes (276 – 195 BCE): Greek mathematician, astronomer and geographer who used the Sun to measure the size of the round Earth, getting very close to the figure we use now.

Hipparchus (190 – 129 BCE): The first to notice that the positions of the equinoxes were changing, or, in other words, he discovered precession, which we'll discuss later.

Claudius Ptolemy (90 – 169 CE): Greek astronomer and mathematician who came up with a mathematical model of the solar system which had the Sun, stars, and planets revolving around the Earth. Known as the Ptolemaic system, it remained in place for hundreds of years, although of course, it turned out to be wrong. However, this system was very accurate at predicting where stars and planets would be at any given time, and Ptolemy also used that to come up with horoscopes.

Abd al-Rahman al-Sufi (903 – 986 CE): Known as Azophi to westerners, he was a Persian astronomer who made the first known observation of a group of stars outside of the Milky Way, and that group is what we now call the Andromeda galaxy, although he had no way of knowing just how far away they were.

Nicholas Copernicus (1473 – 1543 CE): On his deathbed, Copernicus published his idea about our universe being heliocentric (revolving around the Sun). He proposed that the Sun was near the center of the universe, and the Earth spun daily on its axis while revolving annually around the Sun. His proposal did not sit well with the beliefs of the Catholic Church, which is why he waited so long to publish them. Copernicus was also a healer and an astrologer and became the transition-point between the ancient astronomy of Aristotle and Ptolemy (that centered mainly on the position of heavenly bodies and saw the heavens as largely unchanging) and the modern astronomy of Galileo and Newton (which looks in detail at heavenly bodies and finds the cosmos to be very dynamic).

Tycho Brahe (1546 – 1601 CE): Danish astronomer who created precise astronomical measurements for over 700 stars and built an observatory, named Castle Uranienborg after the Greek muse of astrology, Urania.

Galileo Galilei (1564 – 1642 CE): Italian astronomer who brought the refracting telescope to astronomy and used it to observe stars within the Milky Way, the phases of Venus, craters on the Moon, and the four largest satellites around Jupiter (now known as the Galilean moons). He defended the idea of the planets revolving around the Sun, and he wound up under house arrest at the end of his life because of it.

Johannes Kepler (1571 – 1630 CE): Using detailed

measurements of the paths of planets kept by Danish astronomer Tycho Brahe, Kepler determined that planets traveled around the Sun in ellipses rather than in circles. From that insight, he derived three laws related to the motions of planets that astronomers still use today.

Giovanni Cassini (1625 – 1712 CE): Italian astronomer who measured how long it took the planets Jupiter and Mars to rotate and discovered four moons of Saturn and the most visible gap in the planet's rings.

Christiaan Huygens (1629 – 1695 CE): Dutch scientist who proposed one of the earliest theories about the nature of light, a phenomenon that had puzzled scientists for hundreds of years. His improvements on the telescope allowed him to make clear observations of Saturn's rings and to discover its biggest moon, Titan.

Sir Isaac Newton (1643 – 1727 CE): Building on the work of those who had gone before him, this English astronomer is most famous for his work on forces, especially gravity. He calculated three laws describing the relationships between the motion of an object and the forces acting on it, known today as Newton's laws of motion.

Edmond Halley (1656 – 1742 CE): A British scientist and astronomer who didn't discover Halley's comet, but he reviewed historical comet sightings and proposed that the comets which appeared in 1456, 1531, 1607, and 1682 were all the same one, and predicted that it would return in 1758. Although he died before then, he was proven correct, and the comet was named in his honor.

Charles Messier (1730 – 1817 CE): French astronomer who compiled a database of objects known at the time as "nebulae" which included 103 objects at its final publication, although additional objects were later added based on his personal notes. Many of these objects are often listed with their catalog name, such as M31 (Messier object 31) for the Andromeda Galaxy. Messier also discovered 13 comets over the course of his lifetime, and his catalog of nebulae was compiled to help him identify which objects were *not* new comets.

William Herschel (1738 – 1822 CE): British astronomer who cataloged over 2,500 deep sky objects. He also discovered Uranus and its two brightest moons, two of Saturn's moons, and the Martian ice caps. William trained his sister, Caroline Herschel (1750-1848), in astronomy, and she became the first woman to discover a comet, identifying several over the course of her lifetime.

Maria Mitchell (1818 – 1889 CE): The first woman to be given an astronomy-related award, which was for the discovery of a comet now known as Miss Mitchell's Comet. She did that using a three-inch refracting telescope.

Henri Poincare (1854 – 1912 CE): He proposed the idea of gravitational waves in 1905.

Williamina Fleming (1857 – 1911 CE): Scottish astronomer who discovered the first white dwarf star and many other celestial objects. She also worked on the first Henry Draper Catalog listing the spectral classifications of over 10,000 stars down to nineth magnitude[ii].

Edward Barnard (1857 – 1923 CE): Discovered 17 comets, produced a photographic atlas of the Milky Way, and Barnard's Star is named after him, since he had measured the proper motion of that red dwarf that is relatively close to us in the constellation Ophiuchus.

Henrietta Swann Leavitt (1868 – 1921 CE): She discovered that the brightness of a special star-type known as Cepheid variables was related to how frequently the star pulsed. This relationship allowed astronomers to calculate the distances to stars and galaxies, the size of the Milky Way, and helped in establishing the expansion of the universe. While she got little credit for her work during her lifetime, she now has an asteroid and a crater on the Moon named after her.

George Hale (1868 – 1938 CE): Proved that sunspots were linked to the Sun's magnetic field, and he was also the driving force behind the 60-inch and 100-inch telescopes at the Mount Wilson Observatory and the 200-inch telescope, named the Hale Telescope, at Palomar Observatory.

Albert Einstein (1879 – 1955 CE): This German physicist proposed a new way of looking at the universe based on the ideas that the laws of physics are the same throughout the universe, that the speed of light in a vacuum is the same to everyone, and that space and time are linked in a form known as spacetime, which is distorted by the mass of an object.

Harlow Shapley (1885 – 1972 CE): American astronomer who calculated the size of the Milky Way galaxy and the general location of its center.

Georges Lemaître (1894 – 1966 CE): A Belgian Catholic priest, mathematician, astronomer, and professor of physics who first noticed the relationship between the distance to galaxies and the speed of their movement away from us (as indicated by their redshift).

Fritz Zwicky (1898 – 1974 CE): Swiss astronomer who made many theoretical and observational discoveries including the first identification of the effects of what would become known as dark matter[iii].

Edwin Hubble (1899 – 1953 CE): This American astronomer demonstrated that some objects in the sky existed outside of the Milky Way, so our galaxy wasn't the complete universe. He went on to determine that the universe itself was expanding at speeds proportional to their distance from us, giving us a formula, which later came to be known as Hubble's law, and is now officially called the Hubble-Lemaître Law.

Cecilia Payne-Gaposchkin (1900 – 1979 CE): Her doctoral thesis investigated stellar spectra, leading her to propose that the Sun was composed mainly of hydrogen and helium, and she went on to investigate stellar evolution and variable stars.

Carl Jansky (1905 – 1950 CE): An early pioneer in radio astronomy who discovered radio waves emanating from within the Milky Way galaxy.

Clyde Tombaugh (1906 – 1997 CE): An astronomer from Kansas who started out building his own telescope and ended up discovering our favorite dwarf planet, Pluto, while working at the Lowell Observatory in Flagstaff, AZ. Some of his ashes were carried aboard the first and only (so far) spacecraft to visit Pluto, New Horizons.

Fred Hoyle (1915 – 2001 CE): An English astronomer noted primarily for the theory of stellar nucleosynthesis. He is also known for his controversial stances on other cosmological and scientific matters, such as his belief that the universe was in a steady state and had no beginning. He coined the phrase "Big Bang" to poke fun at the theory of an expanding universe with a definitive starting point, and the name stuck.

Nancy Grace Roman (1925 – 2018 CE): She became NASA's first Chief of Astronomy and served through the 1960s and 1970s, during which time she played a foundational role in the planning of the Hubble Space Telescope[iv].

Vera C. Rubin (1928 – 2016 CE): Her observations of the rotation rates of galaxies demonstrated the existence of dark matter[v].

Frank Drake (1930 – 2022 CE): He was one of the founders of the Search for Extraterrestrial Intelligence (SETI) and came up with the Drake equation, a mathematical formula used to estimate the number of detectable extraterrestrial civilizations in the Milky Way galaxy, although the answer incorporates a lot of assumptions/guesses within its calculations.

Carl Sagan (1934 – 1996 CE): American astronomer who carried out some important scientific studies in the field of planetary science, and also popularized astronomy and educated the public about it.

William K. Hartmann (born 1939): American astronomer who, in 1975, put forth the most widely accepted theory for the formation of the Moon, suggesting that, after a collision with a planet about the size of Mars, debris from the Earth coalesced into the Moon.

Stephen Hawking (1942 – 2018 CE): He had many significant insights related to the field of cosmology, demonstrating that the universe has a beginning and also will likely end, and that the universe has no boundary or border. He also had a lot of ideas about the nature of black holes, including showing that they should emit small amounts of thermal radiation and are therefore not entirely black.

Jocelyn Bell Burnell (born 1943): An astrophysicist from Northern Ireland who was the first person to discover pulsars.

David Reitze (born 1961): Executive director of the LIGO Scientific Collaboration which, in conjunction with the Virgo collaboration, announced the first observations of gravitational waves in February 2016.

Katie Bouman (born 1989): She was part of the team that produced the first image of a black hole, and she was a leading force in the development of the algorithm that enabled the team to combine information from radio telescopes around the world to produce the image.

Part 1: Nigredo (Blackness) - Big Bang

Astronomical Alchemy

Chapter 1: The First Few Minutes

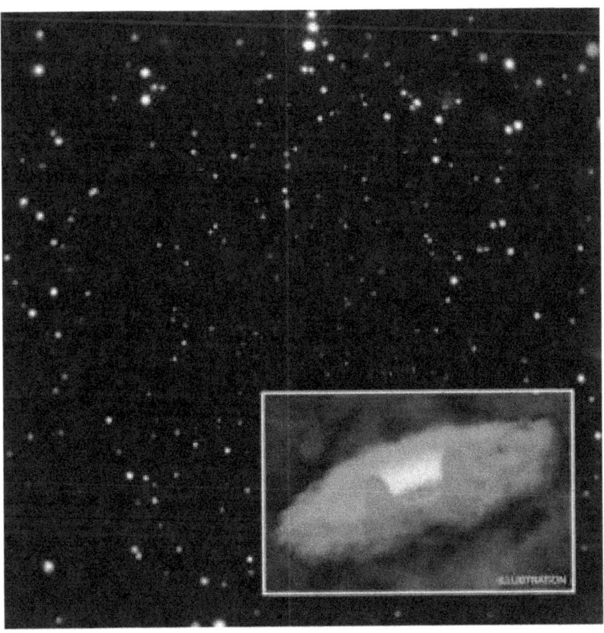

Alchemical Step 1: Nigredo or Darkness

The alchemists recorded that the first step in the recipe for the Philosopher's Stone is Nigredo, meaning blackness. That involved cleansing and cooking the ingredients until it was a uniform black substance. That might not seem to be a good way of describing the Big Bang, because we normally visualize that as a giant explosion consisting of a brilliant and massive flash of light, but maybe that's not the correct image.

We are going beyond normal astronomy to discuss cosmology here, and looking at how big the universe is, looking at the small particles that everything is made up from, and exploring the Big Bang that we imagine started it all. We'll also look at whether there was something before the Big Bang, and question whether there actually was a Big Bang.

The Big Bang

It should be pointed out that the Big Bang may or may not have been big, but it certainly didn't bang, and there was no light (in the normal sense of the word) at all. We don't know how it came about, whether it was all the matter and energy that makes up our universe suddenly emerging from absolutely nothing, or if it emerged as a bounce from a previous collapsing universe or if it was a large quantum fluctuation in an essentially dead universe. All we do know is that it must have started somehow.

It is generally believed that a minute fraction of a second after whatever was the true beginning, a dramatic super-expansion, that is called "inflation", effectively erased all trace of whatever went before, so all we have to start out with is pure speculation that can run wild. We'll talk about inflation some more very soon.

We have no way of seeing events back then, but we can

gain an idea about what happened in those first few minuscule fractions of a second by observing what happens in particle accelerators like the Large Hadron Collider in CERN, Switzerland. They do not get up to the kinds of energy levels that existed around the time of the Big Bang, but they are working their way up in that direction and it gives clues that let us theorize about what occurred at the start of everything[vi]. Perhaps I should say hypothesize, rather than theorize, because it's still very much guesswork, even if those guesses are sometimes calculated.

Initially, we can say that there was just some form of super energy compressed into a very small volume, maybe smaller than a speck of dust. So, it was small not big. Well, it may have been infinite, but we'll get to that. There was no air, and sound results from changes in air pressure, so there was no bang, no noise at all. This compressed energy had no atoms, no electrons, and it had no photons so there was no light. However, as mentioned above, it didn't stay compressed and small. The immensely hot, dense mass of primordial energy expanded extremely rapidly and began cooling as it grew in size and became more spread out.

Although I said there was no light, there certainly was energy and when the universe was big enough (but still sub-microscopically small) that energy would almost certainly have manifested at times as photons, which we tend to equate with light. But visible light consists of photons within a specific range of wavelengths, and these early photons were far more energetic than those we observe today as visible light. Plus, light needs to move as a beam of photons, from one point to another, in order to be observed, and these early photons would have been colliding and combining with other particles before they could go anywhere, and then be emitted again only to be absorbed by another particle.

We tend to visualize the Big Bang as a fiery ball blasting out into the void, but the theory says that the Big Bang created all of time and space, so there really was nothing for it to expand into since the Big Bang was the moment of creation of all matter, energy, space and time. So, we have to say that the Big Bang occurred simultaneously everywhere[vii]. If that is all true, how can we really tell if it did expand or just reorganized itself internally over time? If the universe was infinitely small, i.e., zero size, and it multiplied in size by a billion for every billionth of a second over the period of almost 14 billion years that the universe has existed, then its size would still be zero. Multiply zero by any number, for as many times as you like, and you still get zero. But, however it happened, we now see matter in the form of distant galaxies flying out and away from one another. For now, we will accept that the Big Bang was the start of everything, and we'll look at some of the problems associated with that idea later.

The observable universe may have emerged from an infinitely small spot, but if the universe is really infinite now, then it had to have been infinite in the beginning too, just much (maybe infinitely) denser. In our story, we are at the start of what is sometimes called the "Hot Big Bang" which spans the time that the universe was hot and dense, from year 0 to around year 380,000. That covers about the first 15 minutes of our "cosmic year" where we have reduced the time from the Big Bang to the present to one Earth-year.

The Big Bang happened around 13.8 billion years ago, with a possible range from about 12 billion to 14.5 billion, but it is normally said to fall between 13.7 billion and 13.85 billion. Star HD 140283 (the Methuselah Star, made almost entirely of hydrogen and helium) is estimated to be around 14.46 billion years old +/- 0.8 billion years so, at the low end of its estimated age, it ties up nicely with the normally accepted date of the Big Bang. Light from what was the most distant known galaxy at the time I was writing this, GN-z11, has taken 13.4 billion years to reach us and taking the universe's expansion into account, it is now expected to be 32 billion light-years away[viii].

As we start out, we are dealing with the Planck Time, from time zero to about 10^{-43} seconds (that's a 1 divided by a number starting with 1 and followed by 43 zeros, so it is an unbelievably small number), and we really have a hard time explaining anything here. To get anywhere near an explanation for that initial time period we would need a Theory of Everything (TOE), and that is slow in coming. With the super high temperatures above 100 million million million million million (10^{32}) K initially, the energies of particles in thermal equilibrium were so high that the gravitational force between particles was as strong as the other forces. But what we now know as the observable universe was still only as big as a single subatomic particle.

Regarding temperatures, degrees kelvin (or K) starts at absolute zero and each degree is the same "size" as a degree Celsius (or Centigrade, as I was taught in England). So,

273.15 K is freezing point in kelvin, while -273.15°C is absolute zero in Celsius. For the stupendously high temperatures that we are dealing with here, you can say that Kelvin and Celsius are equivalent, with a ridiculously low margin of error. The actual conversion rate to Fahrenheit from Celsius is "multiply by 9/5 and add 32" which I round off to "multiply by 2 and add 30" for day-to-day weather temperatures. When we are dealing with the kind of temperatures around the period of the Big Bang, they are just unimaginable on any temperature scale, so just think of them as "bloody hot".

Then we get to a time that can be explained somewhat by GUT (the Grand Unified Theory – where gravity has separated out, but the other forces are still combined or indistinguishable). That era lasted to about the 10^{-36} second mark[ix]. In our current universe, electricity and magnetism work together, and we have a theory that brings in the weak force and shows how and when the weak force separated from the electromagnetic force. GUT brings the strong force into the fold, but the theory is still a work in progress, and TOE (Theory Of Everything, bringing gravity into the mix) is still a dream. In the time we have now reached in our story, the universe has still about a trillion times higher energy than that generated by the most powerful collisions in our most advanced particle colliders.

A neutrino today could travel through a light-year-thickness of lead without reacting to any of the atoms along the way (well, about half of a beam of neutrinos would get through without interacting with the lead), but the universe around the time of the Big Bang was so dense that no neutrinos could move far.

Somewhere around this 10^{-36} second point, cosmic inflation may have kicked in, increasing the observable universe by a factor of more than 100 trillion trillion (10^{26}), bringing it to something like the size of a beach ball. Seemingly, the universe is supposed to have had about one gram of a special type of matter that created the force to drive this fantastic expansion, and it then decayed to form the matter and energy we now know in our universe. That would have been at around the 10^{-34} second mark as inflation ended. So, where did all the matter and energy that we see in the universe today come from if there was only one gram of some sort of stuff to start with? One idea is that matter, antimatter and regular energy have positive energy, which is balanced exactly equally by gravity which has negative energy. At least, gravity's energy has to be negative if we want the math to work out right and give us a universe for free. Where dark matter and dark energy fit into that formula is unclear to me, but let's not lose sleep over it for now.

Inflation is felt necessary to account for the consistency seen in the density and temperature of distant parts of the visible universe where light wouldn't have had time to cross and coordinate conditions. Other regions beyond our observable universe may vary more, but our observable universe came from a minuscule point that got stretched, so everything is consistent. In this picture, the universe apparently expanded from a tiny dot of super energy to something around the size of a beachball and, within the blink of an eye, it's the size of Earth, then bigger than the Solar

System, and it was still pure energy. Not that some of that energy didn't try to transfigure itself into elementary particles, but the heat from this surging energy would tear any particles apart before they had a chance to form properly.

At a trillionth of a second (10^{-12}) the universe was still basically pure energy at a density of 10^{35} pounds per cubic foot (10^{36} kg per cubic meter). Any particles that formed would have been changing type constantly, say from a top quark to a photon, then a Higgs boson, everything being in constant flux, recycling particles into energy and vice versa. There could also be much more massive (i.e., high energy) particles that the Grand Unified Theory postulates but that we've never seen because we can't produce sufficiently high energy levels to get close to creating them. Nothing was stable, it was still a writhing mass of intense energy making rather desperate but futile efforts to create particles that instantly disappear or change into something else.

The rapid expansion caused by inflation would have effectively cooled off the universe and spread out any previously existing particles of matter so much that there was virtually no matter to be found in any particular region. Consequently, the universe needed to be reheated and filled with the ordinary matter and dark matter that we see (and don't see in the case of dark matter) today. That supposedly came about by the inflatons, which may have been of many types and that are thought to have driven inflation, using their leftover energy to decay into the quarks, electrons, etc., that make up matter today[x].

By about 10^{-10} seconds, temperatures dropped sufficiently so that top quarks (the most massive known particle) started disappearing from the universe (there wasn't the heat/energy to sustain them), followed very soon by the disappearance of the Higgs bosons, Z bosons and the W bosons. Those latter two mediate the weak nuclear force responsible for some decays of nuclear material in atoms, and the Higgs boson is related to the Higgs field that gives particles the property of mass.

Now the strong nuclear force breaks away from the electroweak force, and at around 1 microsecond (millionth of a second, 10^{-6}), we enter the Quark Era as particles like quarks (that make up protons and neutrons) and gluons (the "glue" that holds an atom's nucleus together) start to emerge from the primordial energy-soup and survive in a quark-gluon plasma. We don't see individual quarks nowadays but, at the point in the universe's evolution that we have reached, the energy or temperature is still way too high to allow them to combine into protons and neutrons. This quark-gluon plasma apparently had a smooth soft texture like water[xi], although it was very far from cool and refreshing. We can recreate a quark-gluon plasma momentarily during collisions of atomic nuclei in our particle accelerators.

The universe may have had a phase transition when the universe cooled to about 3,000 million million K (about 5,500 million million degrees F), at which time the weak and electromagnetic forces were no longer the same strength and started to go their own ways. That probably occurred around the 0.1 nanosecond (10^{-10} seconds) mark, as the electroweak force split into electromagnetism and the weak nuclear force. Also, around that time, something happened to create a distinction between matter and anti-matter. That allowed most of the universe's antimatter and matter to start annihilating with each other, leaving the matter we now enjoy. We've no real solution as to why or how some matter remained, but it seems that the weak nuclear force probably had something to do with it.

Exactly what the role of the electromagnetic field was in the early universe is still unclear, but the search is underway to try to find "primordial" magnetic fields. Simulations appear to indicate that including magnetic fields in models of how the universe evolves leads to values closer to the ones we measure, helping solve the "Hubble tension" (where the universe seems to be expanding faster than the standard model suggests)[xii].

Next, we come to Big Bang Nucleosynthesis as the universe cools enough for familiar particles to form. At about a tenth of a millisecond (10^{-6} of a second), the quarks and gluons come together as protons and neutrons, while the universe continued rapidly cooling and expanding. Now we can say we have the nucleus of hydrogen atoms (one proton each)[xiii].

At $1/100$ (10^{-2}) of a second, the temperature was about a hundred thousand million (10^{11}) K (180 thousand million degrees F) and there were electrons and positrons in large numbers, so hot they acted like radiation, and there was a similar number of the various neutrino types, plus photons of equal number, all created out of pure energy and often annihilating one another again. The density was about 4 thousand million (4 x 10^9) times that of water, so dense that even neutrinos were still constantly colliding with other particles.

Around the 1/10 of a second mark, the temperature was down to 30 thousand million (3 x 10^{10}) K (54 thousand million degrees F), and the energy density was about thirty million times the density of water. At that temperature it was easier for neutrons to turn into protons by ejecting an electron and an electron antineutrino, so there was now about a 38%/62% proportion of neutrons to protons.

At last, the universe passes the 1 second mark, and at about 1.09 seconds the temperature was around 10 thousand million K (18 thousand million degrees F), and neutrinos and antineutrinos begin to behave like free particles, no longer in thermal equilibrium with electrons, positrons and photons. Electrons and positrons were beginning to annihilate more rapidly, and the neutrons/protons proportion was then about 24%/76%.

By about 13.82 seconds, the temperature was a mere 3 thousand million K (5.4 thousand million degrees F), electrons and positrons were annihilating faster than they could be created from photons and neutrinos, and the energy released slowed the temperature drop a bit. Neutrinos were not heated that way, so they dropped to about 8% cooler than the photons, etc. The neutrons/protons ratio was down to about 17%/83%.

Did It Really Bang?

The idea of a starting point in time for the universe did

not sit well with scientists when the idea was first proposed, partly because there seemed to be no good way to start things off without invoking some form of god. That may have been argument enough against the Big Bang idea for some people, but for others the theory may have just seemed too "easy" an explanation. Consequently, we had ideas like the Steady State Theory, which proposed that the universe might be expanding but only because more atoms of hydrogen were being formed constantly in the void between stars and galaxies, letting the universe stay basically the same forever.

Evidence for the Big Bang has grown since the idea was first formulated, but we still have lots of unanswered questions. For instance, we know almost nothing about the dark matter and dark energy that make up 95% of the universe, nor do we understand how ordinary matter wasn't completely annihilated by antimatter[xiv]. And there are probably a lot more questions that we haven't even thought of yet. Appendix B gives a list of some of the unanswered, or not definitively answered, questions that give scientists job security.

Other ideas for getting around the need for a real initial starting point include a cyclic universe with eras of expansion followed by contraction and then a bounce to get things started again. However, any form of cyclical universe can be shown to change over the cycles and to have needed some definite starting point at some time in the long-distant past. Or maybe the Big Bang was a large, or really massive, quantum fluctuation in a previously expanded and fading universe. Anyway, for now we won't worry too much about what happened before the Big Bang itself, especially since the Big Bang is supposed to have created both time and space so, on that basis, there was no "before".

Or maybe there was a "before". There is the idea of "eternal" inflation, where our universe is just one of many bubble universes continually emerging from the inflating field. But even that inflation must have started somewhere, and one suggestion for that is that it was the result of a quantum fluctuation in absolutely nothing. How does absolutely nothing fluctuate in any fashion, quantum or otherwise? It might be more honest to simply admit "We don't know".

Since we don't really know what happened, perhaps I can engage in a little more speculation. We know that the early universe went through a few phase transitions as the different forces separated and went their own way (i.e., gravity became its own thing, then the strong nuclear force did, followed by the weak nuclear force separating from electromagnetism), so what if what we think of as the Big Bang was another phase transition from something that came before. You could say that that sounds like our bubble universe dropping out of "eternal inflation", but I'm thinking of the whole infinite universe (everything that existed back then) transitioning into the possibly infinite universe that we think we could be living in. Naturally, that is wild speculation, and then we have the question of what that previous state was, but it could explain the sudden appearance of an infinite expanding universe.

Atomic Nuclei Form

Getting back to the history of the Big Bang, as the expansion of the universe continues, the energy continues to cool, and soon subatomic particles were able to form and remain stable for at least a while. That didn't really lessen the violence because there were what we think of as normal subatomic particles of matter, plus antimatter subatomic particles (the two form together) and, when those two types collide, they blow up, reverting back to pure energy. But somehow ordinary matter worked its magic and emerged victorious.

Around the 3 minutes mark, the temperature was at 1,000 million K (1,800 million degrees F, which is about 70 times that of the Sun's core), cool enough for the strong force to bring protons and neutrons together to form the first bonded atomic nucleus, deuterium (a form of hydrogen with one proton and one neutron). The neutrons/protons ratio was at 14%/86%, and photons had about a 35% higher temperature than neutrinos. For every 100 seconds that passed, around 10% of the remaining individual neutrons decayed into protons.

Soon we had a hot dynamic plasma of mostly hydrogen and helium nuclei and free electrons, with photons bouncing around between them. Something somewhat similar is happening inside our Sun today, with the result that it takes around a thousand years for a photon to make its way from the core of the Sun to the Sun's surface, but then it takes only about eight minutes for the photon to reach Earth's orbit once it has broken free. However, our growing universe has no "surface" for the photons to break free from, so the photons were still not free to behave like the light waves we know.

By 3 minutes 45 seconds, the temperature was at 900 million (0.9×10^9) K (1,600 million degrees F), letting protons and neutrons form more complex nuclei, including helium. The universe's density was down to that of water, and the universe consisted mostly of photons, neutrinos and antineutrinos, some nuclear material (73% hydrogen, 27% helium), plus the electrons left over from electron-positron annihilation. Just prior to that, the neutron/proton balance was 13%/87% which ties up nicely with the fraction by weight of helium that we observe today.

By 20 minutes, most of the helium nuclei that exist today had formed (light helium consists of two protons and a neutron and is also known as tritium or a triton nucleus; normal helium consists of two protons and two neutrons). There were also some nuclei of deuterium (heavy hydrogen, consisting of a proton and a neutron) and a small amount of lithium (4 protons and 3 neutrons) and beryllium (4 protons and 5 neutrons).

The temperature dropped to about 10 million kelvin (18 million degrees F, similar to that found at the core of a star and actually cooler than the Sun's core) and production of light elements had ceased. Helium ended up being about 25% of all matter by mass. Calculations of how much of each element would be created tie up almost exactly with what we see in fairly pristine cosmic gas clouds that have had little contamination from stars. The observable universe was now light-years across, but still dark because the milky soup of loose electrons and occasional atomic nuclei didn't let light travel far.

At this stage, we are still only talking of the nuclei of elements forming, not full atoms with their electron "shells"; the temperature of the universe would remain far too high to allow that to happen for around another 380,000 years.

How Big is the Universe?

At one time we thought that our galaxy was the total universe, and it is certainly big, taking light about 130,000 years to cross it (estimates range from around 100,000 to 200,000 light-years) and it takes around 230 million years for our Sun to orbit once around it. But we now know that our galaxy, the Milky Way, is not even particularly big compared to some other galaxies, and we estimate that there are about two trillion galaxies in the observable universe, but that number is decreasing as galaxies grow by merging with others.

From observations we know that the universe is expanding, on large scales it is nearly homogenous (has the same average density of matter and radiation everywhere) and is isotropic (it has the same expansion rate in all directions). However, we see maybe only a small part of the full universe, and farther out things could change. We can see galaxies that are now more than 30 billion light-years away, even though their current light could never reach us because they would now be moving away from us faster than the speed of light.

The further out we look (which is also looking back in time) the faster the galaxies are moving away from us, getting close to the speed of light relative to us. Of course, from their point of view, they would be stationary, and we would be flying away from them at close to the speed of light. Even further out, the universe is expanding faster than the speed of light relative to us, so light from any galaxies that far away will never reach us now. These galaxies have slipped beyond our cosmological horizon.

Even though the universe is 13.8 billion years old, the light from very distant parts has not had time to reach us and, in fact, it will never reach us due to the expansion of the universe. That light may be heading in our direction, but it will be getting further away from us because the cumulative expansion rate is faster than light speed over those distances.

Determining the expansion rate of the universe is difficult because measuring distances to galaxies is problematic. The distances to celestial objects are measured by a number of methods. Within the solar system, things like direct laser measurement or observing eclipses can give us accurate distances. Beyond that, we can use parallax, triangulating on a nearby star's observed position against more distant objects when viewed from opposite sides of Earth's orbit. Those distant objects are, for all practical purposes, assumed to be fixed.

Beyond that, we have to rely on so-called "standard candles", such as Cepheid variable stars whose fade/brighten cycle period indicates their intrinsic brightness, so we can compare apparent brightness to intrinsic brightness to give us such a star's distance. Cepheid variables brighten and dim in a predictable way, as Henrietta Swan Leavitt discovered. A Cepheid that's intrinsically brighter has slow, gradual pulsations, while one that's dimmer pulsates more quickly. Using parallax on nearby Cepheids gave a reliable basis for calculating intrinsic brightness against pulsation rate and comparing that with visible brightness gives distance. But we can really only make out Cepheid variables in the Milky Way and in nearby galaxies.

Another distance measuring gauge is type 1a supernovae that all explode with around the same energy (and we've learned how to adjust for the slight differences), so they have similar intrinsic brightness and can be another "standard candle" that is brighter than Cepheids. You need to measure how the explosion peaks and dims in order to get a good sense of the energy output and therefore how bright they really are. Red supergiants at the top of their size range, all seem to have near enough the same intrinsic brightness, so they make another useful "standard candle". Now astronomers are looking into the possibility of using gamma ray burst as another means of measuring distance to galaxies[xv].

Finally, we have the redshift of the spectrum compared to distance that shows the rate of expansion of the universe, letting us extrapolate that the distance to the edge of the observable universe is about 46.5 billion light-years.

Each of these methods of measuring distance lets us measure further and further out, and the overlaps between methods let us correlate the measurements. For instance, in one galaxy we may see Cepheid variables and type Ia supernovae and be able to compare the distance calculations, so we can put more faith in the measurement of more distant galaxies where we can no longer see Cepheids clearly.

We believe that the observable universe is about 93 billion light-year across, roughly 10^{26} meters (100 million billion km). The question is, how far does the whole universe actually go? If there is a cosmologist on a planet in another galaxy, he will see a similar observable universe which may contain our galaxy or maybe not if their galaxy is far enough away. There could be any number, perhaps an infinite number, of "observable universes" out beyond the furthest point that we can see with our most powerful telescopes.

Some estimate the actual size of the universe as being 23 trillion light-years in diameter, some suggest that it is at least 500 times the size of the observable universe which would make it about 46 trillion light-years, while others say it is infinite[xvi]. From observations and measurements, we can see no reason why it might not go on to infinity.

So, the next question is, since stars and galaxies are physical finite objects, can there be an infinite number of them? Can you have an infinite number of finite objects? You won't find an answer to that question here, although it's a puzzle that has bugged me since I was around age 10.

Assuming the universe is finite and that everything

emerged from a minute point, and assuming that we are not so far away from where that point of origin was that it is now way outside of our visible universe, then we should be able to recognize which direction the origin point is. Since everything would be expanding outward from a single point, the further away from that point of origin that we look, the more spread out everything would be, and looking back towards that origin we'd see matter more densely packed together. Yet we don't see that. Sometimes the constant expansion that we do see, wherever we look, is visualized as being like looking at points on the surface of a balloon all moving away from one another at the same rate as the balloon is inflated. So, it is suggested that our universe is like that balloon, but instead of the two-dimensional surface of the balloon being inflated in three-dimensions, we are a 3 (spacial) dimensional universe inflating in four spacial dimensions[xvii]. The fact that we are in a 3-D universe seems to be very fortunate because, if our understanding of the math involved is correct, having a different number of spacial dimensions (say 4-D) would mean that things like orbital mechanics wouldn't work, so you couldn't build solar systems or galaxies.

We can't see the motion of galaxies directly, but we can measure their redshift, a sort of Doppler shift. Light from objects moving towards us will shift to a higher frequency (maybe even going to gamma rays from the original visible light), while light from objects moving away from us will be shifted to a lower frequency (maybe down to radio waves). We don't just look at the color because stars vary in color anyway, but we look for distinctive absorption lines that result from light interacting with elements (such as hydrogen) and at where those lines fall on the observed spectrum. The relationship between distance and redshift was predicted by Georges Lemaître and confirmed by Hubble's measurements, so that relationship of redshift to distance is now officially called the Hubble-Lemaître Law. The redshift also tells us how far back in time we are viewing, and how old the universe was when light left its source galaxy. A redshift of 0 is the present day, 1 is about seven billion years ago, 6 is when the universe was only a billion or so years old, and infinity would be "time zero" or the Big Bang itself.

The Hubble radius is the distance from us beyond which galaxies are moving away faster than the speed of light, and that's around 14 billion light-years. An object at that distance has a redshift around 1.5, meaning that the light wave and the universe itself has stretched to two and a half times its original length from when the light was first emitted. The relative-size-increase factor is 1 plus the redshift value. We've seen supernovae out to redshift 4, the most distant galaxies seen have a redshift of about 11 and the CMB has a redshift around 1,100. So, how can we see light from objects that are now moving away from us faster than the speed of light? Basically, it is because the light that we are seeing left them a long time ago, before the galaxies were receding at that speed relative to us, but the universe has continued expanding as the light made its way to us, increasing the speed of recession.

Although the universal expansion is now increasing, the reverse used to be the case. With galaxies far enough away from us, when the light left them, the universe was still slowing and so their light could catch us up and entered our Hubble radius from outside. The Hubble radius grows over time and lets us see objects that were previously outside of it. But eventually, no galaxies outside of our Local Group will be visible because the Hubble radius is growing so slowly that the universal expansion will pull previously visible objects outside it.

How Fast is the Universe Expanding?

It was 1929 when Edwin Hubble demonstrated that the universe was expanding, but the rate of expansion (Hubble's constant) has been difficult to measure. Hubble himself overestimated the expansion rate by a factor of 7; in the 1990's it was normally quoted as between 50 to 100 km/s/Mpc (that's kilometers per second per megaparsec and a Mpc is 3.26 million light-years). One way to measure the rate of expansion is to directly measure how fast galaxies are moving away from us at various distances using Cepheid variables and type Ia supernovae (exploding white dwarfs) imaged by the Hubble Space Telescope, and that gives expansion rates of between 72 and 76 km/s/Mpc. Let's call it 73 km/s/Mpc which was the figure most recently arrived at using mostly Cepheid variables to gauge the distances[xviii]. You can also infer the Hubble constant from studying the CMB (Cosmic Microwave Background) from maps produced using the Planck telescope data. The Cosmic Microwave Background is the primordial light released when atoms formed about 380 thousand years after the Big Bang. That assessment from the CMB, using the formulas derived from the standard model of cosmology, shows the Hubble constant as being about 67 km/s/Mpc. Why the mismatch between the two ways of calculating the expansion? There are a number of ideas banded about, but nothing that is fully accepted so far. It may just mean that there is some misunderstanding in the calculation of one or both figures, and that trying different ways of measuring the expansion will resolve the issue. Others think that it is indicating that there's a fundamental error in our understanding of how the universe works[xix].

Another method of measuring the distance to galaxies, as mentioned earlier, is by using stars that are known as TRGBs (Tip of the Red Giant Branch) and using those results gives an expansion rate that nearly splits the difference between the other methods. The figure that Wendy Freedman and her team came up with using this method was 69.8 km/s/Mpc. TRGBs are old stars found mostly in the relatively dust-free outer regions of a galaxy, and it is thought that interference from dust may have affected the interpretation of some Cepheid variables used for establishing distance, which are more commonly found near the center of galaxies. Cosmic dust can mess up observations, as the folks at the BICEP2 observatory in Antarctica can tell you. More measurements are needed to confirm or rule out this idea, but an agreed upon rate of expansion for the universe seems a bit more likely now[xx].

Although the expansion rate is referred to as the "Hubble constant", it isn't constant but changes over time. Anyway, to visualize what sort of expansion we are talking about, a one-meter (three-foot) ruler would expand about the width of an atom in one year if it was subject to universal expansion[xxi]. However, objects that we see around us, including rulers, are held together by the electromagnetic force and nuclear forces and are not subject to the universal expansion. Neither are things like galaxies that are held together by gravity.

What Shape is the Universe?

Einstein saw gravity not as a force but as a property of spacetime with, as John Wheeler said, the curvature of spacetime telling matter how to move and matter telling spacetime how to curve, which is a bit of a chicken-and-egg situation. Karl Schwarzschild derived a solution to General Relativity's equations for spacetime around what we now call a black hole, and the result can be pictured as a curved funnel shape dropping down from an otherwise flat "sheet" of spacetime and ending in a singularity at the bottom. That solution to the equation also came up with what is known as the Schwarzschild radius, or event horizon, beyond which even light and information cannot escape from the black hole. Other physicists looked at the equations and saw the possibility of there being different overall "curvatures" to spacetime, one having positive curvature (normally pictured as a sphere or an inflated balloon), another having negative curvature (pictured as a saddle-shape), and the third as being flat[xxii]. If you imagine a triangle drawn on each of these three surfaces, the sum of the internal angles of the triangle on the positively curved surface would add up to more than 180°, on a negatively curved surface it would be less than 180°, and on a flat surface we'd have traditional Euclidean geometry with the internal angles adding to exactly 180° or two right angles as Euclid would have expressed it. Knowing the distance that an acoustic (sound) wave would have travelled in 380,000 years gives a baseline for triangulation, and a way of checking on the curvature of spacetime against the CMB. The results seem to indicate that the universe is flat. It's normally thought that the universe may not have started out that way, but that inflation smoothed out any preexisting curvature.

When we talk of the geometry of the cosmos, we are not referring to its external shape but to an internal property. For instance, it tells how laser beams would behave as they shoot out across the universe. If two such parallel laser beams start to curve in towards each other, then we have a closed universe (positively curved). If they pull apart, then the universe is open (negative curvature). If they stay parallel, then it's flat. Whether the universe is open, closed, or flat depends basically on the density of the universe and the rate at which it is expanding. For all of those options, by themselves, present-day expansion would be slowing down due to gravity.

Before Dark Energy was discovered, it was thought that the overall "shape" of spacetime would predict how the universe would end. If the gravity from all matter was stronger than the expansion, everything would get pulled back together and we would be living in a "closed" or spherical universe and we would be headed for a future

"Big Crunch". If the driver for expansion was stronger than the counteracting gravity, the universe would perpetually expand, and we would be living in an "open" or saddle-like universe and were headed for a cold and lonely end as expansion slowed but never fully stopped. So, astronomers and cosmologists were trying to establish if we were above or below the "critical density" that is the balance-point between continued but slowly decreasing expansion and a Big Crunch. That "critical density" turned out to be equivalent to around six hydrogen atoms per cubic meter or, say, six grains of sand in a volume equivalent to the Earth[xxiii]. Measurements seem to indicate that we are at the balance point between a "closed" and an "open" universe and live in a "flat" universe. Now that we know of Dark Energy, the current best explanation is that we live in a flat universe that will, nevertheless, expand forever at an increasing rate because of that dark energy.

Chapter 2: Inflation and Multiverses

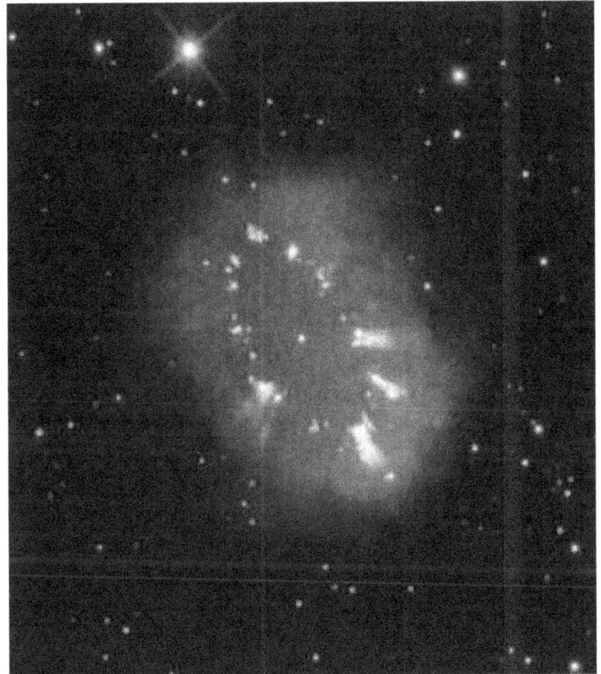

Over the past century or so, our view of the size of the universe has grown from just being our Milky Way galaxy to a universe that might be infinite, then came an idea that there could be other separate universes, and now we are getting multiple versions of the multiverse.

Theories or hypotheses that lead to multiverses include

inflation, string theory, and quantum mechanics, and just for fun we have antimatter adding extra universes too. Each of these versions of the multiverse might have an infinite number of universes, which could all be of infinite size. Feeling small?

Inflation and Multiverse Type 1

In order to make sense of the homogenous and isotropic universe that we see, cosmologists imagine that soon after the initial event of the Big Bang there was a brief and spectacular period of hyper-fast expansion – known as cosmic inflation – about which we know next to nothing. That supposedly took the universe suddenly from being around a billionth the size of a proton to the size something like a soap bubble or even a beachball, or maybe bigger, depending on who is describing it. That idea came about because it was noticed that the universe is remarkably consistent in its temperature and average density. It was also noted that spacetime is totally "flat" as far as can be ascertained[xxiv]. That "flatness" means that our normal rules of geometry apply and also meant that the universe was not placed in danger of collapsing soon after it appeared. To account for that smoothness, Alan Guth and others proposed the idea of the universe having an extremely rapid expansion almost immediately after it came into existence, and that expansion would dramatically stretch and even out any inconsistencies or "lumps". In that way, the universe was forced to be consistent and "flat".

It seems that what first got Alan Guth's attention was

actually a different problem, namely the fact that we don't see magnetic monopoles around today. These are particles that would have only one magnetic charge, so they would either be north-charged monopoles or south-charged, unlike magnets we see these days that have north and south ends. These magnetic monopoles require the kind of energy that was only available immediately following the Big Bang in order to be created, but they also would be stable, and so they should be plentiful today. Alan Guth's inflation would spread out all the magnetic monopoles so much that the odds of coming across even one today is effectively zero. Problem solved, and it was found that inflation could also ensure that our universe was homogenous and flat, so three problems solved for the price of one[xxv]. Nowadays, there are doubts as to whether such monopoles ever existed, although they are consistent with string theory and the grand unified theory (GUT), but, whether they exist or not, inflation is still maintained as useful to the homogeneity and flatness issues. If magnetic monopoles did exist in the early universe, perhaps their magnetic attraction pulled them together into large clumps that then collapsed under their gravitational attraction and formed the cores of the supermassive black holes that seem to have arrived on the scene way earlier than expected.

Anyway, inflation was supposedly driven by some form of "inflaton field" (which would be a type of field known as a scalar field) that then converted into normal matter once it had finished driving the rapid expansion. Having spread any existing matter out so thinly as to be almost non-existent, the universe needed to be repopulated with matter and, having the inflaton field convert itself into the matter we know, solves that problem. Also, the energy density of the universe would then be too low to create magnetic monopoles (if they ever existed) and so that ties up with the fact that we don't see them today.

Alan Guth's team wasn't alone with the inflation idea, and a number of people had come up with similar hypotheses about a rapid inflationary period, including one such expansion that was driven by the Higgs field, which is the only scalar field that we have demonstrated really exists.

While the idea of inflation is consistent with the universe that we see, my question would be: Why shouldn't the universe be consistent? Personally, I have more problem seeing how enough inconsistencies arose to enable stars and galaxies to form than I do in wondering why the universe as a whole seems so consistent. If it came into existence as a minuscule bundle of energy (or even as an infinite but instant eruption of energy) that started pushing outward, then I have to wonder why there should be any "lumpiness" to begin with. Any particles that formed were immediately bumping into each other and forming new ones, energy was being shared and constantly transformed and everything would have come into thermal equilibrium. Light couldn't travel anywhere but was merely pushing outward against everything else immediately around and that alone should have levelled temperatures out, even if they weren't the same all along, and could have been a driver for the rapid expansion at the same time.

Of course, "pushing outward" is probably not the right way to phrase it anyway, because there was supposedly nothing for it to push out into. Our brains are very much limited in their capability to comprehend what was going on (mine is anyway), because it's not often that we see universes being created, even if some physicists think it's happening all the time.

The inflation hypothesis suggests that this super-rapid expansion was driven by a kind of energy field that was subject to quantum fluctuations that are explained by the Heisenberg Uncertainty Principle. Tiny quantum fluctuations then got stretched out to large enough regions to become the slight hills and valleys that we see in the 0.002% temperature and density variations of the CMB. So, these quantum fluctuations got stretched by inflation from microscopic to macroscopic scales and, in turn, resulted in the gravitational variations that transfigured the gas into today's galaxies and cosmic large-scale structure[xxvi].

The next question is: If inflation did occur, why did this rapid expansion stop? What turned off the inflaton field? Some people believe that it didn't stop, and that our universe simply dropped off as an independent bubble universe as that portion of the field collapsed and our universe then continued to expand at its own pace, while inflation went on its merry way, spinning off other bubble universes. And to make matters more interesting, the laws of physics could be completely different in those other universes. So, we could have an infinite number of universes, each one potentially of infinite size. If you have a lump of radioactive material, you know how much of it will have decayed within a set period of time, based on its half-life, but you can't make a specific prediction for when a particular atom would decay. It is thought that inflation might work in a similar fashion. The rapid expansion is going on, and then one area of it "decays" into a more stable state and becomes a "pocket universe", while the rest goes on expanding with more and more "pocket universes" dropping off over time.

Some people see such "eternal inflation" as having been going on forever, but studies of such inflation and other potential "eternal universes" all seem to indicate that there must have been a starting point sometime[xxvii]. To account for inflation, or at least the universe, having started somehow, quantum physics' uncertainty principle is called on to somehow allow a universe or inflating bubble to suddenly emerge out of nothing. Now we need to explain where the uncertainty principle got started, but maybe philosophers are better at that kind of thing.

String theories suggest that there is a vast landscape of universes with at least 10^{500} different possibilities for how physics might work in the myriad bubbles of the inflationary multiverse, each with different laws and different values for the constants of nature. So, maybe it is that the multiple "pocket universes" being created by inflation make up these universes that string theory suggests.

Observing primordial gravitational waves, which would have a much longer wavelength than those that LIGO detects, but which ESA's planned space-based Laser Interferometer Space Antenna (LISA) could see, might tell us if inflation really happened and whether the universe is actually flat or not. People have tried finding indications of the gravitational waves that inflation should have created in the image of the CMB, but have failed to find them, leading to more doubts about inflation[xxviii].

Multiverse Type 2 and Quantum Mechanics

If that inflationary multiverse isn't bizarre enough, some quantum physicists believe that when a quantum event occurs, since these events are believed to be probabilistic rather than deterministic, then all possibilities occur and the universe splits with one probability happening in each. And yet scientists supposedly don't believe in magic! Even so, this kind of multiverse is not totally unbelievable, although I'll argue that it seems to violate some of the laws of physics. Matter is popping into existence and vanishing again in the near total vacuum of space all the time, and matter is really just some form of "condensed" energy, so let's give this quantum multiverse a hearing. My presentation here of the Multiple Worlds interpretation of quantum mechanics is rather simplified, and there are in fact multiple versions of the Multiple Worlds interpretation. In some, the multiple universes are "real", in others they only appear to be real.

In quantum theory we have a wavefunction, determined by the Schrodinger equation, moving through an infinite-dimensional place called Hilbert space. In traditional quantum physics, particles are associated with this wavefunction and can be in several places at once (i.e., they are in a superposition), but when you try to observe or measure a particle, its wavefunction collapses and the particle gains a specific location.

Hugh Everett looked at the situation where the Schrodinger wavefunction is all that you have, complete with its superpositions, and that concept led to the idea of the existence of parallel universes where you might live out countless variations of your life.

Technically, these would not really be separate physical universes (that would go against conservation of matter and energy) but each individual version (or superposition) of "you" would be looking out onto a different landscape made up of different options available from the various superpositions of particles and things made up from them. Here, you don't really have matter and energy, except as portrayed by the evolving wavefunction. In this interpretation of quantum physics, there is no collapse of the wavefunction, no "real" particles, only different views looking through the one universal wavefunction. Or should that be "multiversal wavefunction"?

Some people do look at this interpretation as saying that the wavefunction actually "splits" every time a quantum event happens, or maybe every time you make a choice, but as mentioned above, that gets you into too much trouble with conservation of matter and energy. If the universe does actually split, it would also break the no-cloning rule of quantum physics in a BIG way. If the universe split, the second version of the universe would have an identical cloned copy of every subatomic particle in the universe with the exception of the one that triggered the split. And if quantum cloning works, goodbye to any idea of quantum encryption. If it didn't split and there's only the wavefunction, with no real particles or atoms, then how does chemistry work? If it's just images projected from the wave, like some sort of Star Trek holonovel, who is writing the script and directing the action?

Multiverse Type 3 - Antiuniverses

Maybe there's one other type of alternative universe, one that may have been observed by the Antarctic Impulse Transient Antenna (ANITA) in 2016 when it detected a high-energy particle blasting out from Earth rather than heading in from space. Two years previous it had seen the same sort of thing. One explanation is that the particle came from a parallel universe created concurrently with our own but made of antimatter and which is traveling backwards in time in relation to us.

A proposal by Latham Boyle, Neil Turok and Kieran Finn suggests that our universe and a time-reversed version of the cosmos emerged from the Big Bang in a way that can be visualized as two cones touching tip to tip. This would be a universe-antiuniverse pair, emerging from nothing, and it makes some specific predictions about the nature of dark matter that might be testable someday.

If such a universe type exists, that means we can double the infinite number of universes we had from inflation and quantum mechanics, because there could be matter and antimatter versions of all of them.

Multiverse Type 4 – Mathematical Complexity

Max Tegmark added his own multiverse variety. The Mathematical Universe Hypothesis is based on the idea that mathematical existence equals physical existence, or that math is at the basis of all reality. Consequently, our universe is based on the mathematical structure that we are used to, but other mathematical structures are theoretically possible.

This type of multiverse has also been thought of from a different viewpoint. Whatever way the universe started, it had a set of laws built into it somehow, and the application of those laws formed the basis for how the universe evolved. But the set of laws that our universe uses may be just one set from multiple options that are in a superposition, and other combinations might give different sets of laws leading to different universes. Since such laws are defined by math, we come back to different mathematical universes again. Such alternative mathematical structures could form the basis for complete sets of nested multiverses, starting off with the multiverses springing from inflation, each of which might have a matter and antimatter universe which in turn are really only probability waves producing infinite views that look to us like individual universes continually branching off from one another.

One's Enough to Handle

All of these multiverse types may be theoretically possible and are well worth considering. They might even be essential to the existence of our habitable universe. However, for now they should be described as hypotheses rather than theories, because there does not appear to be any way to test them, let alone prove that they really exist. Never mind the fact that the term "universe" is supposed to encompass everything there is. One infinite universe is more than enough to deal with in this book, so for now, let's stick with the universe we have. But we'll come up with one or two other multiverses before we get to the end of the book, I promise.

54

Chapter 3 - Observing the Universe in Multiple Ways

Nowadays, we can get information about the cosmos from a number of different sources, including photons, neutrinos, cosmic rays, and gravitational waves[xxix]. All of these are produced by different forces, so even the presence or absence of any of these signals can tell you quite a bit about the object they came from. However, all of them require different kinds of detectors, and astronomers are trying to witness events from as many of these sources as possible to get as big a range of data as they can.

We have started looking for all of these kinds of signals from violent astronomical events and among the things that we have discovered are rare ultrahigh-energy cosmic rays originating from outside our galaxy, and high-energy neutrinos forming a cosmic background that pervades the universe. What the objects are that are sending out those kinds of signals is obscure, but suggestions include super-strings, dark matter, and "defects" in the structure of the universe. There seems to be plenty out there still to be discovered and explained.

Photons

Photons are the traditional signals that we have used to look at the universe and they are fluctuations in the electromagnetic field. Such electromagnetic radiation includes visible light, radio waves, microwaves, infrared (that JWST looks for) to ultraviolet light, X-rays, and gamma rays, the last two of which are blocked by the atmosphere. Gamma rays are the most powerful photons and can be produced here on Earth by nuclear reactions, including fusion, fission, and atomic decay. Since stars are gigantic nuclear furnaces, they are also sources of gamma rays. Gamma-ray bursts (GRBs), which can last from a couple of seconds to a few minutes, are born from the collapse of a massive star or the collision of neutron stars, and a GRB can put out more energy in seconds than the Sun does in billions of years. The shorter gamma ray bursts are likely to be from colliding neutron stars that can be relatively close to us, while the longer GRBs are those from the collapse of the kinds of supermassive and fast spinning stars that were common in the early universe and have therefore travelled very long distances before being picked up by our detectors. Gravitational wave detectors can tell gamma-ray detectors like the Swift satellite roughly where to look for the short bursts, even if gravitational waves don't give exact locations yet. Gamma-ray detectors give good locations and can then tell optical telescopes, including Hubble, where to look. One GRB event ended up being captured in X-ray, optical light, infrared light, radio waves, and gamma rays (gravitational observatories were offline at the time), and the evidence indicated that it was a kilonova from a neutron star colliding with another neutron star or with a black hole.

Neutrinos

Neutrinos are produced via the weak nuclear force, and they have no charge and very little mass. As a consequence of that, they do not like interacting with other matter, which makes them notoriously difficult to detect, although we know that the Sun and other nuclear furnaces produce them.

The first neutrino detector was a huge tank of dry-cleaning fluid situated a mile down in an abandoned gold mine and that set the pattern for neutrino detectors being large fluid-filled voids surrounded by sensors to try and catch the occasional interaction. Since neutrinos only weakly interact with matter, huge vats of water are needed, or people have tried instrumenting the ocean or the Antarctic ice sheet to find them (ice works better for finding the direction of the source object). Neutrinos from the supernova, SN 1987A, in the Large Magellanic Cloud were seen two hours before light from the explosion arrived. Since neutrinos hardly react with matter at all, those formed in the core of a collapsing and exploding giant star will head out immediately at roughly the speed of light, but the shockwave from the collapse can take hours to work its way to the star's surface and blast out as the visible supernova. Neutrinos travel straight, passing directly through objects, and consequently they are easily traceable back to their origin source. The IceCube observatory in Antarctica detected neutrinos from a blazar in 2017. A blazar is a galaxy with a central supermassive black hole shooting out jets of ionized matter at nearly the speed of light. IceCube captured signals coming from the trail of particles by observing a single neutrino's impact that pointed to a source in the constellation Orion; NASA's Fermi Gamma-ray Space Telescope showed that blazar TXS 0506+056 was in that direction, making that blazar the third identified individual source of neutrinos (after the Sun and SN 1987A), and the blazar was observed to be flaring up at that time. Those detections indicated that blazars play a role in generating cosmic rays, which are critical intermediaries in producing those high-energy neutrinos.

The word "neutrino" means "tiny neutral thing" and they appear, at first glance, to be inconsequential. But we are finding that they play a major role in some of the biggest events in the universe, such as supernovae. With their connection to the weak nuclear force, they may also have played a big role in the fact that matter won out over antimatter and that there is anything at all that could create stars, galaxies, and us. It is even thought that a type of neutrino not yet observed might make up a large proportion of the dark matter that is 85% of all matter in the universe and which provides the scaffolding for the cosmic web[xxx].

Cosmic Rays

Cosmic rays are high-energy protons and atomic nuclei left over from matter that has been torn apart as a result of the strong nuclear force and have then been accelerated to near the speed of light. Their collisions along the way with interstellar dust can result in the creation of complex organic molecules, and their interaction with interstellar gas might generate neutrinos. Places like the Pierre Auger Cosmic Ray Observatory use massive water tanks to observe the subatomic shrapnel from cosmic rays hitting the atmosphere. Our Sun produces cosmic rays, boosted by coronal mass ejections and the like, but these are not very powerful compared to others. Cosmic rays originating within our galaxy are probably accelerated by shock waves from supernovae, but very rare ultrahigh-energy cosmic rays have been observed that seem to come from outside our galaxy, possibly from actively feeding supermassive black holes. There seems to be more of these coming from one side of the sky than the other, and that doesn't square too well with the idea of us having a homogenous and isotropic universe. But each discovery seems to raise more questions, and it may have something to do with the heliosphere (our protective shield provided by the solar wind, and which stops a lot of the cosmic rays from outside of the solar system) not being completely spherical because it is deformed by the Sun's movement through the galaxy.

Gravitational Waves

Gravitational waves are produced, fairly obviously, by gravity and were suggested in 1905 by Henri Poincare. Einstein's general relativity equations from 1916 gave a mathematical basis to the idea of gravitational waves, but the waves themselves were not observed until 2015. Einstein's theory indicates that our universe is an interlinked fabric of matter, space and time and, when one part of it (such as matter) moves, the rest of the fabric ripples in response.

The first gravitational wave detected by LIGO arrived at Earth on September 14, 2015, at 9:50 a.m. UTC (Coordinated Universal Time), having travelled for 1.3 billion years from the merging of two black holes each thirty times the mass of the Sun. Now we could feel the vibrations of space itself. The discovery was announced on February 11, 2016, by teams from the Laser Interferometer Gravitational-wave Observatory (LIGO) and Virgo collaborations. From 2015 through 2020, 50 gravitational wave events had been identified by the LIGO and Virgo observatories, resulting from collisions of neutron stars and those of stellar-mass black holes. In January 2020, LIGO spotted its first black hole/neutron star merger, actually seeing two such mergers that month.

The first successful simulation of a binary black hole merger was carried out in 2005 by running Einstein's field equations on a supercomputer. Millions of such simulations have now been run and each unique waveform is stored as a sort of template. When a new signal is picked up by LIGO, Virgo, and the Kamioka Gravitational Wave Detector (KAGRA) it is checked against these templates to figure out what is most likely to have caused the signal[xxxi].

LIGO's detectors send laser pulses down a pair of 2.5-mile (4 km) arms that are perpendicular to each other, the pulses being then reflected back and combined, and changes in the interference pattern are noted[xxxii]. LIGO has a total of 4 arms, 2 in Livingston, Louisiana, and 2 in Hanford, Washington. Virgo's and KAGRA's arms are 1.9 miles (3 km) long. The interferometers can detect changes in the arms down to $1/10,000^{th}$ the width of a proton. The GEO600 detector has 600-meter-long arms, and that is a bit too short to pick up much, but this German-British instrument has pioneered the squeezed-light technique for improving the sensitivity of gravitational wave detectors.

Pulsars spin with millisecond periods and changes in those periods can be detected at the hundreds of nanoseconds level[xxxiii]. As such, pulsars give us many arms and they have been hinting at gravitational waves for decades. Pairs of supermassive black holes slowly spiraling towards each other in the center of galaxies should create long wavelengths in the gravitational field, each wave maybe taking decades to pass us. The strongest evidence for these comes from the North American Nanohertz Observatory (NANOgrav) that uses data from the Green Bank Telescope and the much-missed Arecibo Observatory, looking at records related to 47 pulsars observed since 2004. These pulsars form a galaxy-size gravitational wave detector. As spacetime around Earth is stretched and compressed, the pulsar signals should arrive slightly earlier or later than expected, and these variations should be correlated in a pattern, not just be random. Such delays in the order of hundreds of nanoseconds have been found, but we cannot definitely say if they are due to gravitational waves, although some possible alternative causes have been eliminated. At the time of writing (2021) they still have another two years of data to work through[xxxiv].

The International Pulsar Timing Array (IPTA) is looking for variations in the timing of the pulses across a number of pulsars, trying to listen in on collisions of supermassive black holes from the merger of galaxies. The gravitational waves that IPTA is looking for would have a very low frequency with a period of 1 to 10 years, so long observation times are needed.

Of all the signals, gravitational wave signals can be expected to be the first to arrive, so it will be good to have improved detectors that can pinpoint events that others can also study. The Japanese Kamioka Gravitational Wave Detector has recently begun operations with target sensitivity planned for 2024, there is the ongoing Advanced LIGO-Virgo upgrade, and a new LIGO facility is being built in India. Then there is ESA's plan for the space-borne gravitational wave observatory, the Laser Interferometer Space Antenna, which is scheduled for launch in the early 2030s.

It has been calculated that if gravitational fields are quantum fields, then gravitons should create jitters that could be detected in the gravitational waves being recorded by the current gravitational wave observatories. If that idea pans out, it will show that reconciliation is possible between general relativity and quantum mechanics, which would be a major step forward in our understanding of the universe[xxxv]. Also, while gravitational waves move through space at around the speed of light, massive objects such as black holes or even clouds of dark matter are expected to slow down the part of the signal that encounters the object. That should result in a "gravitational glint" showing up in the signal, that could identify information about the unseen object[xxxvi]. Gravitational wave astronomy is just starting to learn what information it can uncover. For instance, it is hoped that gravitational wave astronomy will be able to show if cosmic inflation really happened, what happens when supermassive black holes collide as a result of galaxy mergers, and whether microscopic primordial black holes could be a component of what we know as dark matter[xxxvii].

Teleparallel Gravity

General relativity sees mass as curving what is called spacetime, and the mathematics used to describe gravity in this framework uses formulae related to curvature. Those formulae have survived the test of time, but Einstein himself wasn't totally comfortable with them and he looked into the concept of using math related to torque, or twisting, instead. This idea became known as teleparallel gravity. Einstein hoped to come up with a theory that would explain both gravity and electromagnetism. Sadly, although the idea worked just as well as his original general relativity formulae for gravity, it still didn't work for electromagnetism.

If you are wondering about the name, it is called "teleparallel" because it looks at how parallel lines behave in space. The idea of teleparallel gravity never went completely away, and it has been tweaked and has become viewed as a possible solution to a number of problems in physics, and it has also been shown to relate to string theory, although with string theory being still very questionable, teleparallel gravity can't depend on that for too much support.

However, versions of teleparallel gravity do away with the need for dark matter and dark energy, with those two simply being the natural outcome of the interaction of mass with this revised version of gravity. Cosmic inflation is also explained as a natural consequence of events at the birth of the universe, without the need of a special "inflaton" field. Using the formulae of teleparallel gravity to interpret the data from the CMB also brings the estimated expansion rate of the universe in line with that derived from measurements of Cepheids and supernovae, making the so-called "Hubble tension" disappear. Sadly, all of these wonderful solutions currently rely on different tweaks to the teleparallel formulae, so there is still work to be done to see if this is a real solution to a multitude of problems or just wishful thinking. But it is a subject worth watching[xxxviii].

The Need for Quantum Gravity

We know something's not quite right with general relativity, but it always seems to check out for just about any situation. However, in an extreme environment, like in the vicinity of a black hole or among the particles at the center of a neutron star, the equations don't seem to work so well. Gravity looks totally different from other forces from a mathematical point of view, and it's also way too weak. The math of general relativity is described as "simple and elegant", but that's if you are deeply into differential geometry, and consequently people revert to Newton's equations if they can and, luckily, that works for most practical situations.

We hope there's a theory of everything that unites gravity with the other forces, and there are promising leads on a Grand Unified Theory uniting the electroweak and the strong nuclear force. Results coming out of the LHC are indicating that we are probably seeing the first clear indication of a connection between the strong nuclear force and particles that have been thought to have no connection with it. These results involve anomalies related to how beauty (or bottom) quarks decay and have led to the idea that there might be something called a leptoquark that is the carrier of a new force that can transform quarks into electrons and their heavier cousins, muons and taus. Nothing definite yet, but watch this space or, better still, watch the results coming out of the upgraded LHC as it gets back into business[xxxix].

But so far, the feebleness of gravity has thwarted efforts to bring it into the fold. It is also true that general relativity doesn't play nicely with quantum physics, so the effort is on to find some way of describing gravity in quantum terms. Quantum gravity theories include string theory and loop quantum gravity, but none of the ideas have been confirmed by particle experiments, and we're not sure if they can be. Let's hope they can, because reconciling the theory of the massive (General Relativity) with the theory of the very small (Quantum Physics) will give us a much better handle on what is going on in the universe.

Chapter 4: Dark Matter and Dark Energy

A Bubbly Universe

The International Space Station orbits about 240 miles above us (about half as far as from London to Edinburgh, or from San Francisco to San Diego) and from there you could see the continents and make out the cities (especially by the city lights on the dark side of the planet) but you couldn't tell where one country ended and another one started. From the orbit of Mars, about 140 million miles (225 million km.) away on average, the Earth would out-shine the stars, but without a telescope it's just a bright blu-ish dot. Out at Saturn's orbit, say 870 million miles (1,400 million km.), Earth is scarcely noticeable unless you know where to look. About 65 light-years away and the Sun is no longer a naked-eye object, so forget about seeing the Earth. If you travelled 200,000 light-years above the galaxy, you could see the spiral arms of the Milky Way, but you'd be hard-pressed to find the Sun, even with a good telescope, and yet we're still less than a twelfth of the way to our near-est galaxy, Andromeda. Move out further and we see that the Milky Way and Andromeda are part of the Local Clus-ter of galaxies, bound together by gravity, and that the Lo-cal Cluster is part of a supercluster. Moving much further out, we see that these superclusters are not separate, iso-lated congregations but are linked to one another by strings of galaxies and clusters that connect together at nodes, forming an interconnected cosmic network with somewhat spherically shaped, tremendously large voids between them. These voids are getting bigger, like bubbles being in-flated. What we are seeing are the effects of dark matter and dark energy as they shape our universe.

By measuring the shape of the universe (see Chapter 1) we have been able to infer the total amount of mass in the cosmos. What had taken us by surprise is that the atoms that were thought to be the building blocks of everything, only make up about 5% of this mass, leaving 95% unexplained.

The missing content is both invisible and able to pass through us undetected. All we can see of it is its effects, and that seems to indicate that it consists of two markedly different substances, called dark matter and dark energy. They seem to be polar opposites, with dark matter pulling things together while dark energy pushes things apart.

Dark Matter

In our solar system, the planets further out from the Sun travel a lot slower than those closer in. For instance, Mercury orbits the Sun at a speed about 106,000 mph (170,600 km/hr.), whereas Neptune orbits at about 12,250 mph (19,700 km/hr.). Astronomers thought that the same would apply to stars in a galaxy, but instead they found that they all seemed to be moving at much the same speed. The stars out near the galaxy's edge seem to be moving too fast for where they are, and the puzzle was why they weren't being thrown out of the galaxy. The only thing that seemed to explain that was to imagine that there was something else out there with several times as much mass as that which we could account for by the total mass of all the stars and planets, etc. It was speculated at first that the extra mass might be something as simple as black holes and other matter that doesn't show up easily, which got the name MACHOs (Massive Compact Halo Objects) but, when that didn't seem to work, we imagined something exotic like WIMPs (Weakly Interacting Massive Particles) which is supposedly some type of particle currently unknown to science. However, none of these ideas seem to be panning out, or at least they have been very difficult to substantiate so far.

Evidence for dark matter comes mostly from the movement of galaxies, where stars far from the center have much higher velocity than predicted from the distribution of visible matter alone, which should seem to have the stars flying out of the galaxy and into the void. Evidence for dark matter also comes from galaxy clusters which wouldn't hold together without it. Gravitational lensing shows it as well, distorting light from background galaxies into a ring of mirages. We can see the effects of dark matter closer to home too. In the Milky Way, the "bar" extending from the galaxy's core seems to be slowing by about 13% every billion years, with that slowing apparently being caused by dark matter[xl]. So, we know it's there, but we don't know what it is. The current list of candidates for dark matter includes WIMPs, axions, or something like these but, whatever dark matter is, it doesn't interact with electromagnetic radiation which makes it very difficult to detect.

Ignoring any contribution from dark energy for now, dark matter is estimated to make up about 85% of the matter in the universe, so normal matter made of regular protons, neutrons, and electrons makes up only about 15%, yet it's only that small amount that we see and really know anything about. The only thing we know for certain about dark matter is that it does not appreciably radiate, reflect or absorb light, but it is gravitationally attractive. Some of the dark matter might be primordial black holes that formed soon after the Big Bang, but there must be some other exotic form of matter too.

MOND, or MOdified Newtonian Dynamics, involves a revision to how gravity works, and it was considered to be an option for explaining dark matter, and still is by some, but a study of merging galaxies in the early universe showed a merger rate between 5 and 10 times greater than that predicted by MOND. The dark matter particles seem to act to decelerate the galaxies as they pass through each other making them more likely to slow down and merge[xli], rather than simply pass through each other and continue on their merry way. The MOND model works well at explaining the motion, etc., within individual galaxies, but it has problems with larger structures and with interactions between galaxies.

So, as other options got ruled out, WIMPs became the best guess almost by default. Such particles seemed as if they might be fairly easy to detect, but the most sophisticated detectors that we have come up with have failed to do so. It is estimated that vast quantities of WIMPs could have been produced during the first millionth of a second or so after the Big Bang, and they should have reached a state of equilibrium with the surrounding plasma of quarks, gluons and other subatomic particles. It has been calculated that if dark matter particles interact through a force that is about as powerful as the weak nuclear force, then the number of particles from the Big Bang would roughly match the abundance of dark matter whose effects we see in the universe today.

Supersymmetry (SUSY) adds to the standard model of particle physics by suggesting that every particle in the standard model has a supersymmetric partner. The supersymmetric version has an 's' added to the front of the particle's name, so the supersymmetric version of quarks and electrons are called squarks and selectrons. Why SUSY concerns us here is that the lightest supersymmetric particle (LSP) might be a candidate for dark matter and might be the looked-for WIMP. There is, also, a hypothetical fourth type of neutrino, called the sterile neutrino, and they would interact only via gravity. However, an experiment carried out at the Fermi National Accelerator Laboratory failed to find any indication of sterile neutrinos, so they may have to be ruled out[xlii].

Axions are a more recent idea for a dark matter particle[xliii], having a tiny mass (say, around a billionth that of an electron) but packed tightly together throughout the universe. If they exist, axions might also solve the CP (charge parity) violation issue that has left us (thankfully) with more matter than antimatter. It was thought that a hidden force and its associated field might be the cause of the matter-antimatter imbalance and axions were envisioned as emerging from such a field[xliv]. Axions have been searched for using powerful magnetic fields in an attempt to convert them into photons, but these experiments have yet to place much in the way of constraints on the properties of the particles or show that they really exist.

It has been noticed that the normal idea of cold dark matter works well at explaining the movement of galaxies in a galaxy cluster but is not so good at explaining the rotational speeds of stars in a galaxy. In that latter case, some additional influence seems to be needed. It's as if two types of dark matter are required. One idea to solve that involves dark matter being something like axions that would already be packed fairly close together, and the gravity involved with a galaxy would result in them being packed even more densely. That, together with the extremely cold temperatures of space could make the dark matter become a superfluid which would then provide the additional influence needed to hold galaxies together. Galaxy clusters are spread out too much to generate the gravity needed to turn the dark matter into a superfluid, so you get the same dark matter behaving differently in different situations[xlv]. A neat solution, but we still need to establish what dark matter really is in order to see if it is the right idea.

It has been proposed that there is a fifth force (in addition to the electromagnetic, weak, strong, and gravitational forces) and it has been called X17. This was postulated to explain the trajectory of certain particles during nuclear transitions, and there would also be an X17 particle, a boson that has been described as a dark photon but with a mass about 16.7 MeV[xlvi]. If it exists, the X17 particle might weakly communicate forces between dark matter particles, but there is limited evidence for such a force. In fact, none of these options for dark matter have been proven, one way or another, yet. Evidence has been mounting to show that dark matter exists, but very little to explain what it is. Another suggestion has been that dark matter is ordinary matter that has fallen into a fourth spacial dimension, so we can't see it but only feel its gravitational influence[xlvii]. Of course, that's probably even more difficult to test and prove than the other ideas.

If all these ideas about what constitutes dark matter are coming up blank, how about another alternate universe – a mirror universe? This is not the Mirror Universe with the evil Captain Kirk, but one resulting from a parity violation with some elementary particles. That idea has been suggested as a solution to why the lifetime of neutrons not incorporated within an atomic nucleus differs slightly depending on which of two methods you use to record the survival time. This mismatch could be indicating matter that we cannot observe, but which could be forming its own stars, galaxies and life, and would only interact with us via gravity and might be what we call dark matter[xlviii]. Some researchers think that they have a way to prove if such a mirror universe exists, but nobody is expecting quick results.

We can't leave string theory out either. String theory suggests that elementary particles are lengths or loops of "string" and the various vibrations that the strings have will define what type of particle it is. This would be like the different strings on a harp that each have their own vibrational frequency producing a different note. It has been suggested that dark matter is made up from strings the same way, but the strings are vibrating one "octave" higher than those forming ordinary matter. So, again we might have the same kind of menagerie of particles and forces as with ordinary matter, but in a form that ordinary matter doesn't interact with except via gravity[xlix].

The Dark Energy Survey (DES) analyzed images of 100 million galaxies to see how many images were being distorted by gravitational lensing. That analysis was aimed at producing a map of dark matter throughout the universe and covered a quarter of the southern hemisphere's sky. It showed a similar web-like structure to that seen with galaxies and clusters, with large voids in between[1] which seems to confirm the idea that dark matter was very much involved in the shaping of the universe. There also appears to be more detail in the dark matter web than in the visible web that it is supporting, which might indicate that the real action is happening in the parts that we can't see. However, analysis of the data appears to indicate that matter is spread out a few percentage points more evenly than is predicted by applying general relativity to a model of the early universe, leading to more questions about how well we understand how gravity works.

Dark Energy

With the universe having started in a Big Bang, people thought it would have begun with a rapid expansion, but then gravity would slow everything down until we might see the universe collapsing back on itself, bringing the universe to an end in a Big Crunch. However, as the speeds of objects in the universe were measured, while looking back over time, it was found that the universe did indeed start out by slowing down but, at some point, the slowing stopped, and the expansion started speeding up again. We call the force that is driving this acceleration in the expansion "Dark Energy".

The commonly accepted story is that the Big Bang occurred 13.8 billion years ago, 10^{-32} seconds later inflation ballooned it out faster than the speed of light, and then the expansion continued at a more sensible pace and gravity started slowing it down. Now we have discovered that, around five to six billion years ago, the repulsive force of dark energy began to overpower the attractive force of gravity and increase the expansion rate again.

In the late 1990s, using type Ia supernovae as "standard candles" to gauge the distance to galaxies, and comparing those results to the redshifts, showed that galaxies were not only moving away from us, but they were also accelerating (the reverse of what astronomers had been expecting). That led to the conclusion that normal matter (stars, planets, us, etc.) only made up 5% of the universe, dark matter makes up 27% and dark energy is the remaining 68%. With dark energy we can have a flat universe that also has its expansion continuing and accelerating.

It is only in those past few billion years that dark energy has come to dominate the universe. It is as if there hadn't been enough space for the dark energy, as a property of empty space, to do much. Then, around five billion years ago, matter became disbursed enough for dark energy to overcome gravity in the growing voids and cause the increase in expansion to become noticeable.

We don't really know what either dark matter or dark energy is, and all we can really be sure of is their effects. Calling it "Dark" Energy is really only saying "We don't have a clue". Dark energy has been viewed as being a fifth fundamental force (along with the strong and weak forces, electromagnetism, and gravity)[li], or as an intrinsic energy of space itself. In the latter case, it is often referred to as the "cosmological constant" – Einstein's fudge factor. Dark energy fits the math of the "cosmological constant" that Einstein added to his general relativity theory, but that doesn't explain what it is. The first idea was that dark energy derived from the vacuum energy of empty space that is believed to have virtual particles quantum-fluctuating in and out of existence, but that idea gave a number that was 120 orders of magnitude larger than what was needed to match the measurements given for dark energy.

One of the things that we don't know about dark energy is if it really is a constant or whether it evolves over time. If we consider it as evolving, then dark energy is often called quintessence, after the "fifth element" of the Middle Ages. We don't know what turned off cosmic inflation, so maybe something similar to it turned back on again. If it was to turn off once more, we could still have a Big Crunch, where gravity reverses the universe's expansion. If it did work like the inflaton field, then turning it off would cause the energy to decay into matter, meaning more gravity and an accelerated Big Crunch.

Dark energy appears to stay constant per unit volume but the space between galaxies increases, with the result that the contribution of dark energy continues growing as the universe ages. Fred Hoyle's "Steady State Theory" (which was an alternative at one time to the Big Bang Theory) suggested that atoms of hydrogen were constantly popping into existence in the void of space, enabling the universe to retain the same density of matter despite the expansion of the universe. Now we seem to have dark energy constantly popping into existence in the void to drive the expansion while keeping the density of that energy constant. This expansion is likely to lead to our Local Group of galaxies merging into one galaxy under the pull of gravity, but all other galaxies will end up being pushed outside our field of vision.

Dark energy doesn't affect atoms (the electromagnetic and nuclear forces overpower it), planets (electromagnetic and gravitational forces hold them together), or galaxies and clusters (gravity holds them). Nevertheless, cosmologists can imagine "big rip" universes where dark energy is stronger than local forces and prevents anything interesting (maybe even atoms) from forming[lii].

Sigma-8

Temperature fluctuations in the CMB let us estimate the abundance of dark matter needed to explain motions in galaxies and galaxy clusters, and that shows how dark matter has to outweigh normal matter by more than 5 to 1, and we seem to have roughly 5% ordinary matter, 27% dark matter, and 68% dark energy. A puzzle has emerged from measurements of the extent to which galaxies clump together on a scale of 8 kiloparsecs. That figure is referred to as sigma-8 and it depends on the mass in the universe and the consequent gravity pulling clusters together. Based on established ratios for different kinds of matter and the behavior of gravity according to General Relativity, sigma-8 should be 0.81. Trying to physically measure it using results from observations of weak gravitational lensing gives a value of 0.74, which suggests that there is less matter in the universe than predicted by the standard model. Observing the extent that light from the CMB is distorted by weak gravitational lensing leads to results that might indicate a closed universe with more dark matter, but that seems to go against the speeding up of expansion due to dark energy. If only the universe would behave in accordance with our math, it would be much more convenient but perhaps not as interesting.

Chapter 5: The Dark Ages up to the CMB

Atoms and What They Are Made Of

The universe is big, and the Greek philosopher, Democritus, around 440 BCE talked of planets around other stars, some planets having two suns, etc., all of which we are now discovering to be true with our space-based and other advanced telescopes. But Democritus, and his teacher, Leucippus, also considered the very small, and they are the first known people to talk of the smallest things that all matter is made out of, which they called atoms. What we now call atoms are not quite the indivisible objects that Democritus imagined, but fairly close, and each of us has about 7 billion billion billion atoms in us, more than the number of stars in the known universe. Scientists have now built electron microscopes, such as PICO, that are about a thousand times more powerful than a traditional electron microscope and can actually show the atoms and their alignment in materials.

PICO shows atoms as little balls of stuff, but that is more like an illustration or visualization. Imagine an atom blown up to 10 miles in diameter and, in that case, the nucleus in the middle of the atom would be about the size of a tennis ball and there would be a few electrons buzzing around far from it. Sometimes the electrons are thought of as being like little objects orbiting the nucleus, like planets orbiting the Sun, but it seems that they are better described as being waves of energy forming shells around the nucleus, with the outer shell being at the 10-mile diameter outer face of our imaginary massive atom, and it's that outer shell that PICO "sees".

So, most of an atom can be thought of as empty space, but with this 10-mile size atom, the tennis-ball size nucleus would still look solid. Nevertheless, the nucleus is actually made up of a collection of protons and neutrons. We have developed particle accelerators, like the Large Hadron Collider, that fire protons or other particles at each other and see what bits break off. From these experiments, we have learned that protons and neutrons are made up of even smaller particles called quarks. And just to make things a bit murkier, the "empty space" between the nucleus and the electrons is actually filled with fields, such as the electromagnetic field.

That leaves us with the basic particles of matter being normally thought of as quarks and electrons, but it has been discovered that, under certain circumstances, electrons can be split into three quasiparticles, knowns as spinons, orbitons, holons, which travel off in different directions through material when the temperature is near absolute zero. Each holds one of the electron's properties, specifically spin, orbital location, and charge. Quite what does constitute "elementary" particles seems to be still up in the air a bit.

Anyway, quarks and electrons are about as small as we have been able to go, while there is speculation about even smaller particles. There is an old saying that goes "Big fleas have little fleas upon their backs to bite them; little fleas have littler fleas, and so ad infinitum". Particle physics follows the same sort of logic. It has been suggested that maybe everything is made of little strings of energy, vibrating at different rates and appearing to us as different particles. Since we know that energy and matter are interchangeable, and that matter is a sort of "condensed energy", the idea of subatomic particles being vibrating strings of energy does seem quite feasible. These strings would be extremely small, maybe a millionth of a millionth of the size of a nucleus, or even smaller.

The interesting thing is that the universe is supposed to

have started out as something about as small as a string when the Big Bang occurred. So, perhaps the smallest constituent of matter contains everything needed to make a universe, much like the DNA in every one of our cells contains the information to build us.

The Higgs Field

Some people are concerned that a powerful enough collision might trigger a universe-destroying quantum event called vacuum decay, that we'll discuss later, and the discovery of the Higgs boson in 2012 made that possibility seem more real. That discovery confirmed the concept of the Higgs field, and it's really the Higgs field, not the boson, that plays a central role in particle physics and the nature of the cosmos. The boson is a localized excitation of that field, while it's the Higgs field itself that gives mass to some particles. The Higgs field is thought to have been instrumental in the process called "electroweak symmetry breaking" that separated the weak force from electricity and magnetism. That electroweak symmetry breaking occurred when the universe was about one-tenth of a nanosecond old.

When the Higgs field underwent that transition, it also gave some particles (but not photons or gluons) the ability to interact with the Higgs field, determining the particle's mass and making them move more slowly. Massive particles moving through a vacuum require a push to get them started, while massless ones have no option but to travel at the speed of light. The Higgs field also determines several fundamental constants of nature, including the charge of the electron. The background physical state of the universe is called the "Higgs vacuum" or simply the "vacuum state".

CMB, the Dark Ages, and the Coming of Light

We were just talking about atoms a little earlier, but the point we have reached in the life of the universe is still before atoms existed. We have seen the nuclei of hydrogen, helium, and lithium form, and there were lots of electrons flying about, but the temperature or energy of the universe was still too high to let those nuclei pull in the electrons and form real atoms. Energy in the form of photons was everywhere, but the universe was still very dense and so those photons were constantly colliding with the free electrons and being absorbed by them, and so light couldn't get anywhere. The basic constituents of matter formed in the first fraction of a second after the Big Bang and the protons and neutrons followed soon afterward, with the result that the creation of the first atomic nuclei was completed within the first twenty minutes (and that's real time, not our "cosmic year"). That left the universe as a hot ionized plasma[liii] consisting of those atomic nuclei and the free electrons that were jiggling around rapidly in the fierce energy field.

Once the temperature dropped to about 4,000 kelvin (about 6,700°F or 3,700°C) protons and electrons could combine to form atoms. Although this is the first time that the nuclei and electrons had come together to form actual atoms, this period is known as "recombination" rather than "combination". But the universe was still young, at only 378,000 years old, so we shouldn't fault it for its grammar. In a way, everything was combined as energy initially, then settled out into electrons and quarks (with the quarks then combining to form protons and neutrons), and you could say that the protons, neutrons, and electrons then recombined but forming atoms this time, not energy. However, it seems that the name "recombination" came from the idea of a cyclical universe, with atoms breaking apart as the previous iteration of the universe collapsed into a singularity, and then the components of the atoms recombined at this point in the universe's next go around. Anyway, now we had atoms of hydrogen and deuterium (a heavy isotope of hydrogen) and they made up about 75% of matter. Helium isotopes made up most of the remaining 25%, along with lithium that accounted for about 0.0000001% of normal matter.

At around 380,000 years after the Big Bang, with the temperature down to about 3,000 kelvin (about 4,900°F or 2,700°C), an event called photon decoupling occurred. This was related to recombination and allowed the photons to travel freely. That was actually the event that generated the Cosmic Microwave Background (CMB) radiation, although at that time the radiation was mainly in the infrared portion of the spectrum and became stretched or red-shifted into the microwave band over time[liv]. Now the universe became transparent to radiation, and light could finally break out if there weren't all those atoms in the way. We've reached the end of the Hot Big Bang era and the universe is about 15 minutes old on the timescale of our "cosmic year". If you could look on this dense fog of hydrogen from the outside (but there was no "outside") you would see it glowing red because the temperature would be about the same as the surface of a red dwarf now.

The accidental discovery of the CMB by Penzias and Wilson confirmed the hypothesis that the early universe was one big inferno, glowing with heat. As measurements of the CMB improved, it was shown to correspond with blackbody radiation of 2.725 kelvins (-455 deg F or -270 deg C), exactly as predicted taking into account the expansion and subsequent cooling of the universe. Only minute deviations in that temperature can be detected, with "blotches" (around the size of the full moon as viewed from Earth) of one "color" grouping together in some regions and more spread out elsewhere. These variations are due to very, very slight changes in density, and deviate from the average by no more than 1 part in 100,000. Those variations grew from tiny blips over millennia to become galaxy clusters. Stars formed within those first galaxies and those galaxies gathered into clusters and became the expansive network we see now as a cosmic web.

This era that we are talking about now is called the "surface of last scattering" as it's the end of the Hot Big Bang and a surface in time beyond which photons go from being trapped in plasma to traveling "long" distances. However, this is also the start of the cosmic Dark Ages, when gas is slowly cooling and condensing into clumps, drawn in by the tiny density blips set up by those initial fluctuations.

Once the early atoms had formed, the universe was no longer a plasma soup but became one big mass of fog that lasted for about 200 million years or longer, and it was probably not until about a billion years after the Big Bang occurred that light could really move freely across the universe. This fog of neutral hydrogen would absorb photons and eject them in different directions, only to have them be absorbed by the electrons in other nearby atoms.

But things were happening in the fog. As we've seen, it wasn't completely smooth, and the denser areas of fog (helped by the unseen Dark Matter that makes up over five times as much mass as ordinary matter) started to collapse as gravity began pulling in the matter around it. Opposing that collapse was the universal expansion trying to spread matter out, but gravity is persistent and far-reaching, even if it is weak, and those regions that were denser than others continued to pull in the surrounding gas before it could escape. Having so much more dark matter than ordinary matter, it was easier for the dark matter to start clumping together first, and then its gravity pulled the ordinary matter along for the ride. That speeded things up and allowed galaxies to form earlier than they otherwise would. The forces of gravity, electromagnetism, and the strong and weak nuclear forces have all separated out by this time. The strong nuclear force was holding the protons and neutrons together, the weak force was involved with whether such particles remained in an atomic nucleus or not, and the electromagnetic force was holding the electrons in place to form atoms.

Variations in the CMB show that the density across the universe at that time only deviated by around 1 part in 100,000, but those minute variations in the universe's structure, along with the hidden dark matter, let things happen. Gravity may be weak, but it is persistent and far reaching, and the gravity from denser volumes of the gas and dark matter started to pull more stuff in. Within that dense gas, hydrogen atoms were pushed together and collided to form hydrogen molecules, and those are better absorbers and emitters of infrared radiation (a.k.a. heat) letting the temperature drop. That temperature drop meant that the clumps of gas could contract and compress further. As the mass grew in these regions, the force of gravity was felt stronger and stronger until these regions pulled in such vast amounts of gas that the atoms in the center got crushed together and started fusing once things had heated up to around 1,300°F (700°C). The ionized core of hydrogen is a positively charged proton, and the electromagnetic force would be trying to push them apart as gravity attempted to push them together. As the mass increased, finally the protons got close enough for the strong nuclear force to reach out and pull them together, fusing the protons into a helium nucleus[lv]. A process called quantum tunneling also helped in the bringing together of the nuclei to allow them to fuse. That fusion released a lot of energy, and the energy pushed back against the inward pull of gravity and supported the mass of the collapsing cloud of gas, and we had a star.

Energy in the form of electromagnetic radiation (which includes light), started to escape from what was by then a supermassive star. But its light didn't get far in the fog, although it did start to open up the area around it. We have now reached the period of reionization. The energetic photons from these supermassive and super-hot stars were powerful enough to free the electrons from some of the surrounding hydrogen gas or, in other words, to ionize the gas. Doing that let the photons travel more freely. So, the radiation started pushing out on the gas around the star, burning away the fog and creating room for light to travel.

The first stars probably formed when the universe was around 100 million years old (January 3 in our "cosmic year") and are called, counterintuitively, Population III stars[lvi]. We're not sure exactly when the first stars appeared, and figures from 30 million years to 200 million years after the Big Bang have been quoted, but 100 million sounds like a good compromise. These first stars grew much more massive than they could today, maybe several hundred times the mass of the Sun, or even 1,000 solar masses and shining a million times as brightly. Being so massive, their gravity would have reached out, tugging on other stars and pulling them into the first galaxies over time. There would have been no dust, just the three lightest elements. Surface temperatures of these stars were maybe fifteen to twenty times that of the Sun, and the stars may have lived only a few million years or less. Some believe that the supernova explosions from these first stars left nothing behind, blasting all the elements that they had created into space. Perhaps more likely is that the universe found that it had some of its first black holes. Anyway, the force of the explosions allowed heavier elements to form in the escaping gas as the blast from the explosion compressed the gas and fired neutrons at it.

The darkness was disappearing, and the universe was ready to move on to the next alchemical stage.

Part 2: Albedo (Whiteness) - Stars & Galaxies

Astronomical Alchemy

94

Chapter 6: Stars & Galaxies

Alchemical Step 2: Albedo or Whiteness

The second step in the alchemical process is albedo, or whiteness, which represents purification and approaching enlightenment. For the universe, it was certainly enlightening, because now light was streaming out from the first stars.

The universe prior to that had been an expanding cloud of mostly hydrogen gas until, over 100 million years after the Big Bang, some regions became denser than others and started sucking the surrounding gas in. There was a lot of gas available to pile onto the forming stars. That meant that these early stars became truly massive (hundreds or thousands of times the size of our Sun), and massive stars mean lots of pressure at the center, which leads to a very active nuclear furnace at their cores.

The First Stars

In amongst the radiation, hydrogen gas, and other primordial elements was something we call dark matter, and dark matter's gravitational pull was probably vital in forming the fluctuations we see in the CMB. Dark matter is really more transparent than dark, and it doesn't seem to react with matter by any means other than gravity. Ordinary matter exhibits the electromagnetic force which could help overcome matter's angular momentum as gravity pulled it together, but matter also experiences electrostatic repulsion which held the particles apart. Dark matter isn't influenced by the electromagnetic force, so it was able to clump together first without immediately bouncing back. Baryonic (normal) matter then got drawn in by the dark matter to form minihaloes, which are clumps of gas about a million times more massive than the Sun. These became nurseries for the first generation of stars, and fairly rapidly evolved into galaxies. The cosmic web is still scaffolded by dark matter clumps and filaments. Around the one hundred-million-year mark, one of those clumps became so dense it ignited into a star, and Cosmic Dawn had begun.

The intense starlight ionized the surrounding gas and broke the hydrogen atoms back into free electrons and protons, creating giant bubbles of ionized hydrogen gas around galaxies and we have entered the Epoch of Reionization that completed about the one-billion-year mark (late on January 27 in our "cosmic year"). The intense radiation from the early black holes that will have been trying to feast on the universe-spanning gas cloud, will also have been playing a role in the reionization process[lvii]. Things then continued much the same way, but with a few twists and turns, for almost 13 billion years up to the present era.

The first stars (Population III) were massive, and burnt hot and fast, and used up their fuel rapidly. Then they blew up in a supernova or hypernova, and the remaining body of the star probably collapsed into a black hole. These first stars may not just have been supermassive stars made of normal matter. With six times as much dark matter around as ordinary matter there is an idea that the first Population III stars were so-called dark stars. Dark stars are not dark and might have been the biggest and brightest stars that the universe has seen[lviii], but they also contained dark matter. A leading theory for what dark matter is, is that it consists of WIMPs (weakly interacting massive particles). We know that each particle of ordinary matter has an identical, oppositely charged particle called an antiparticle. Having no charge, WIMPs would be their own antiparticles. When normal particles and antiparticles meet, they destroy one another in a shower of light, energy, and possibly lighter particles, and it is thought that these WIMPs (being their own antiparticles) would destroy each other the same way.

In the denser conditions back then, it is imagined that these WIMPs danced around, annihilating each other in energetic blasts. To me, it seems that these theoretical dark matter particles would have to have some difference between the matter and antimatter versions, some dark property that gets reversed, or else what is there that would react so that they would destroy each other? But assuming that they would annihilate themselves somehow, these WIMPs may have been swept up by the gravity of giant gas clouds that then condensed to form the early stars. The annihilation of the dark matter at the center of the stars would have given an outward push that stopped the star from reaching the density required for nuclear fusion to start. In the dense early universe, such stars might have begun with something like a solar mass, but quickly grow to be a million times larger as more material was pulled in. Being heated differently from "normal" stars, they could grow much larger, becoming puffy giants, reaching out to the equivalent of Saturn's orbit. They are imagined as having been mostly hydrogen, with only about a thousandth part of the star being dark matter, but that could still power the star for millions or billions of years. As the last WIMPs annihilate, there's nothing to counteract gravity, and the stars start contracting. Smaller dark stars (say, only a hundred times the Sun's mass) would start nuclear fusion and begin sustaining themselves as regular stars for a bit longer.

Eventually, all dark stars would have collapsed into black holes, accounting for the early formation of supermassive black holes (many are found within the first billion years of the universe's birth). Dark stars would be relatively cool, only around 10,000 kelvins (17,500°F, 9,700°C) and would radiate at longer wavelengths than warmer objects. They are also expected to pulsate over periods of weeks, although the universe's expansion would have stretched that to a hundred days or more from our viewpoint.

The earliest galaxy so far seen (by early 2022) is known as HD1 and dates from only 330 million years after the Big Bang. Taking into account the expansion of the universe, it is seen to be 33.4 billion light-years away and yet it is unexpectedly bright and hot. One explanation for the brightness is that it contains a lot of the massive and hot first-generation stars, and another is that it has a surprisingly large supermassive black hole for so early in the universe's life[lix]. But surprises don't seem to be that unusual as we explore the early universe, and here we are at about January 9 in the "cosmic year". The preceding part of this paragraph was written before the James Webb Space Telescope sent back its first images, and of course, it had to find an even earlier galaxy in one of its initial images. That is the galaxy known as GLASS-z13 which is seen as it was only 300 million years after the Big Bang[lx].

As mentioned in the previous section, the initial reactions at the core of regular stars involves fusing hydrogen into helium but, before long, the hydrogen in the center of these early stars began to get used up and the now plentiful helium started to fuse into carbon, followed by nitrogen, then oxygen. The elements were getting heavier at each stage and producing more energy and heat to overcome the pressure of the gravity trying to crush the rest of the star down onto this violent chemical-producing core with its raging nuclear furnace. Some of the outer layers of the star got blown off as the star heated up from this fusing of heavier elements.

Then the larger atoms that had already formed started to get crushed together to form iron, and everything changed rapidly. Up to that point, the fusing of the lighter atoms to make heavier ones had been producing vast amounts of energy to push back against the force of gravity and keep these supermassive stars inflated. The action of fusing atoms to create iron did not release energy, but instead absorbed it. Suddenly, there was no longer an outward push of energy to counterbalance the inward pressure from gravity, and gravity had won. The immense mass of this first-generation star collapsed onto its core and then bounced in a tremendous hypernova explosion (a hypernova has be defined by some as a supernova on steroids).

Up to that point, most of the atoms that had formed in the star had been locked up in layers around the star's core, but that explosion blasted vast quantities out into the universe. These first supermassive stars began sending out flashes of brilliant light and other radiation as their short lives came to an end, providing enough energy to create some of the larger atoms beyond iron. The remnant that was left behind, where the star used to be, might have been a black hole or a neutron star. Some of those black holes might have got drawn together to become the centerpiece of galaxies that began forming, and the stars in those galaxies would include some of the heavier elements created by the first stars. The neutron stars come in useful in another way, as we'll see later.

The explosions from supernovae and hypernovae, or maybe from dying dark stars, would have spread heavier elements into the universe, and those blasts also compressed areas of the surrounding gas, leading to more stars being formed. The more stars there were, the more areas of fog got cleared, and by about 200 million years after the Big Bang, starlight was able to beam out reasonably unimpeded across the universe. We still have some of that hydrogen fog left, but most of the space between the stars is clear.

Later Stars

Population III stars were discussed in the previous chapter, and these were the first-generation stars, made almost exclusively of hydrogen and helium with virtually no metals. Population II stars are metal-poor, but they gained some heavier elements from the debris left over from the explosion of Population III stars. These Population II stars are the older stars that we expect to find in a galaxy's halo and its central bulge. Population I stars are the more recent ones, being metal-rich and are what we expect to find in the flattish disk of a galaxy, and our Sun is in that category. Perhaps I should mention that, to an astronomer, "metal" is any element heavier than helium (i.e., anything that needed at least a star to make it, rather than just the Big Bang).

Stars come in all shapes and sizes, such as the red dwarfs that are relatively cool and extremely long lived, the medium sized stars like our Sun which burn fairly steadily for a long time before blossoming up into red giants as the hydrogen fuel in their core gets used up and they start fusing larger atoms. Then they later collapse back into a white dwarf. We also have supergiant stars that live fast and die young.

Our Sun is the nearest star to us and is the independent sort, but many stars come in pairs or larger groupings. 1.3 million binary star systems have been detected by the Gaia space telescope within about 3,000 light-year of Earth[lxi].

Stars are normally categorized by their spectral characteristics and their temperature. In that regard, from hottest to coolest, we have spectral classes O, B, A, F, G, K, M and L[lxii], which you might remember from the first letters of the not strictly politically correct mnemonic: "Oh Be A Fine Girl/Guy, Kiss Me Lovingly". Each class of star is then subdivided by the numbers 0 to 9, again from hottest to coolest. O-, B- and A-type stars appear intrinsically blue to bluish white because they are the hottest group, F- and G-type stars have a faint to rich yellow hue, and K-, M- and L-type stars appear orange to reddish orange because they are relatively cool. In common parlance, red is hot, and blue is cool, but with stars you reverse that tradition.

How bright a star appears in the sky depends on its magnitude. The stars that are normally taken to be among the brightest are defined as magnitude 1, although the brightest star in the night sky (Sirius) is given a magnitude of -1.33. That is because, as the magnitude value goes up, the apparent brightness goes down and the dimmest stars you can see, on a really dark night with no light pollution around, will be magnitude 6. The apparent magnitude of a star depends on its distance from us and its absolute magnitude. The absolute magnitude is defined as what the star's magnitude would appear like from a standard distance of 10 parsecs (about 32.6 light-years).

Stars are born in clouds of gas called nebulae, consisting mostly of hydrogen and helium, with some other gasses often included in the mix. The atoms in these stellar nurseries are normally totally ionized (excited and energized by hot stars within and around them) which causes them to glow. Such clouds also contain dust from previous supernovae. The denser regions start condensing until they collapse to form a protostar. When a critical density is reached in the mass, fusion begins, and a star is born. Other material gets pulled into a rotating, flattened disk around the forming star. Material from the cloud is still falling onto the protostar but, like black holes, such protostars can be messy eaters, and superheated matter gets flung far into space. One such baby star has been observed blowing out gas and dust for around a 10-light-year distance at speeds of over 93 miles (150 km) per second. Bright shock waves, that are called Herbig-Haro objects, have been observed where the outflow strikes the surrounding gas in the nebula[lxiii].

Most nebulae are simply clouds of gas that are not energized or glowing of their own accord, but they may be reflecting light from nearby bright stars, and others are truly dark, consisting of dusty, black grains obstructing light from beyond them. Some glowing nebulae are the remnants, or the outer layers of gas blown off of a red giant star, shells of softly glowing gas that expands outward, belched away by the dying remains of their progenitor star. Others may be the remains of a star's outer layers after it has gone supernova, and that gas will glow for a short while before dissipating.

The Blue Ring Nebula was formed when two stars collided and merged a few thousand years ago, and the "rings" are actually two cones of material that were thrown out but are now dissipating into space[lxiv]. It used to be thought that the expelled gas would be ejected symmetrically around the star, but it seems that it frequently gets ejected in two opposite directions, forming what looks like two cones of material with the star at the position where their narrow ends come together. Presumably, this is driven by the star's magnetic field, similar to how a black hole ejects material that is surrounding it.

One pair of stars, known as V Sagittae, has its stars on a spiraling collision course that are expected to coalesce and go nova around the year 2083 +/- 10 years, at which time it will briefly become the brightest star in the night sky[lxv]. Nebulae can be some of the most spectacular objects seen during your nighttime viewing.

Red dwarfs and mid-sized stars like our Sun, begin and end their lives in much the same way. A few million years after their birth, the central core gets hot enough to support a sustained nuclear reaction, fusing hydrogen into helium to produce energy. Red dwarfs are more active in relation to CMEs (mass coronal ejections) than Sun-like stars are, but both types of stars are more active in that regard in their younger years than they are later in life. But aren't any young ones always the most active? A young Sun-like star was seen blasting out a massive cloud of plasma as a CME, far larger than any such ejection that we have observed from the Sun, thankfully[lxvi].

You might think that smaller stars would use up their fuel quickly and fade away, but red dwarfs (like Proxima Centauri, our next closest star), that might be about a quarter of the Sun's mass, burn their hydrogen fuel slowly and are very efficient at using it, so their lifespan can be longer than the current age of the universe. Consequently, red dwarfs conserve their limited fuel, making them last for trillions of years, while solar-type stars use up their hydrogen in about ten billion years[lxvii]. The term "main-sequence stars" can include solar type stars and red dwarfs, basically any star that is still getting most of its energy from fusing hydrogen. Even brown dwarfs, during the period when they are fusing deuterium (heavy hydrogen) are sometimes included in the category of "main-sequence".

Red dwarf stars make up about 75% of the stars in our galaxy, but we can't see any of them with the naked eye because they are too faint. Solar-type stars, like our Sun, make up about 20% of the stars, and the rest are the massive giant stars that burn hot and use up their fuel fast, then collapse under gravity's pull and explosively rebound as a supernova. With the notable exception of hydrogen, which came from the Big Bang, almost all of the atoms in your body would have been blasted out into space by one of these supernovae. What gets left behind after a supernova explosion might be a neutron star or stellar-sized black hole.

Before they go supernova, massive stars will usually

blow off some of their outer layers of gas. In late 2019, Betelgeuse (the red giant at Orion's right shoulder – the left side as we look at it) started dimming and by mid-February 2020 it was only one-third of its normal brightness, but it is now back to normal. It is believed that it "sneezed" out a cloud of hot gas from its photosphere in the autumn of 2019 and when the cloud was millions of miles from the star it cooled and condensed into dust grains that made the star look dimmer[lxviii], but now (2022) the cloud has spread out and Betelgeuse is back to its normal brightness level. Similarly, red hypergiant star VY Canis Majoris, 4,000 light-years away is being wracked by pulsations and is expelling mass in a series of dusty knots and arcs, some of which date from up to 1,000 years ago to as recent as the year 1990, and the star is known to have dimmed from our perspective around those times[lxix].

The density of dark matter today is insufficient to power stars, except maybe near the center of galaxies. In that kind of location, there could be "WIMP burners", consisting of white dwarfs and neutron stars that have captured enough dark matter to at least partially fuel them and act like a stellar youth serum, keeping them abnormally hot.

Galaxies aren't always kind to their denizens. We have seen stars that have been thrown from the Milky Way at around 1.5 million mph (2.4 million km/hr.). Hundreds of hypervelocity rogue stars have been observed on the outskirts of the Milky Way heading for the Andromeda Galaxy. There's also indication of stars being ejected during mergers of galaxies. In 1977, Hubble imaged hundreds of red giant stars in the Virgo galaxy cluster far from any galaxy. Globular clusters have been seen in the gaps between galaxies in the Virgo cluster, and it's thought that their stars might be orbiting homeless black holes flung from galaxies.

White Dwarfs

As the hydrogen is used up, solar-type stars grow into red giants, getting brighter, bigger, and somewhat cooler, and they will then be mostly helium but with oxygen and carbon in their core. The carbon is probably compressed into diamond. Red dwarfs don't grow in size as they age, but they become a bit brighter and hotter (appearing blue), and they haven't got enough mass to provide the pressure and heat to burn helium into carbon and oxygen. But when both types have fused all the fuel they can, they both become white dwarfs, with solar types blowing off much of their outer layers and leaving the white dwarf as mostly carbon and oxygen. White dwarfs that are the remains of red dwarfs would be mostly helium, but both types of white dwarfs only shine from the residual energy left over from their lives as active stars. Actually, since red dwarfs are so long-lived, the only white dwarfs we are currently likely to see will be those that are the remains of Sun-like stars.

White dwarfs are dead stars, with no nuclear furnace to keep them inflated. Instead, they rely on a feature of fermions (such as electrons, neutrons and protons) that doesn't let any two identical fermions occupy the same space. Consequently, it becomes harder and harder to push the particles closer together. This force keeping the fermions apart is called electron degeneracy pressure. The force of gravity squeezing white dwarfs together means that the more mass they have, the smaller they are, because gravity squeezes them tighter. They can get so dense, with ultra-strong gravity, that even the supermassive black holes at the center of galaxies can have a problem pulling them apart[lxx].

White dwarfs may be small and essentially dead, but that doesn't mean they are easy to understand. There is one that is about 90 light-years away, named DES J2147-4035, that has about 70% the mass of our Sun and is the coolest white dwarf known. That implies that it must be about 10 billion years old, and its parent star might have been among the first stars in the universe. So, what is it doing in the disk of our galaxy where you'd expect to find young stars? Stars that are the age of that white dwarf would normally be found in a galaxy's halo[lxxi].

Another white dwarf that is about 130 light-years from us, and is known as ZTF J1901+1458, has a radius only about 400 kilometers larger than that of the Moon, making it probably the smallest white dwarf known. Although small in size, it is estimated to be 1.35 solar masses, bringing it quite close to the mass limit that would cause it to explode as a type Ia supernova, and it probably resulted from the merger of a binary pair of white dwarfs. It also seems to be shrinking, leading to the idea that the pressure inside it might be close to where protons turn into neutrons. So, if it doesn't explode, it could become a neutron star sometime in the next few hundred million years[lxxii].

A study of white dwarfs in the globular cluster M13 produced a surprise. It seems that some white dwarfs retain more hydrogen than expected, and that hydrogen can be burning on or near their surface, rather than in their core, making them look hotter and younger than they really are[lxxiii].

As previously mentioned, fermions (electrons, protons, neutrons, neutrinos and quarks) obey the Pauli exclusion principle which says that no two or more of them can be in the same place and energy state at the same time. The free electrons get so crowded in the white dwarf's core they have to keep jumping to higher and higher energy states to keep from being the same, creating electron degeneracy pressure that halts the star's collapse. But that kind of pressure can only support the star to a certain point. If it gains even more mass by pulling material from a companion star, or if it collides with another white dwarf, it can exceed the so-called Chandrasekhar limit. Then the core temperature increases, carbon begins burning and a thermonuclear explosion rips the star apart completely in an explosion that can outshine its galaxy. That mass limit is 1.44 solar masses. These type Ia supernovae occur, as a rule of thumb, once per galaxy per century, which means that there are plenty out there to see because there are a lot of galaxies. They have allowed the Hubble Constant (the rate of expansion of the universe, which isn't really constant) to be calculated as 74 km/s/Mpc to an accuracy of around 2.4%, although there are still questions about that figure. Such explosions can create impressive amounts of nickel (among other elements) but things like gold are mostly made in neutron star collisions.

Brown Dwarfs

A very common and rather weird star is something like a small red dwarf, or an extra-large Jupiter. These are called brown dwarfs, and they are stars that just failed to ignite the normal nuclear furnace at their core, but their crushing gravity and decay of radioactive elements help generate heat, leaving them glowing a dark cherry red. They can have clouds of iron vapor and they rain down liquid iron. They would start out with at least the mass of 13 Jupiters, allowing them to burn deuterium in their core for a while, then they cool and shrink as they lose radiative pressure. The coolest known brown dwarf is around -10°F (-23°C).

Black Dwarfs

Black dwarfs are the last stage of Sun-like stars. Our Sun has 4.5 billion years to go before it will have puffed up into a red giant as it burns helium as well as hydrogen[lxxiv]. After a further 1 to 2 billion years, it will have exhausted all of its available fuel and it will contract to a white dwarf made of carbon and oxygen while its outer atmosphere drifts away as a planetary nebula. About Earth-size, what used to be our Sun will be about 200,000 times denser than Earth, no longer generating energy but, starting out at around 10,000 kelvin (17,500°F or 9,700°C), it will continue to shine and radiate heat for about 10 trillion years (about 730 times the current age of the universe). Then it will fade into a black dwarf, of which there are currently none in the universe because the universe isn't old enough to have created them yet. Then the neutrons within the black dwarf will start to decay and the protons will do so as well, but over a longer period of time. We're not too sure about protons, but it seems as if they should decay. So, over a very long period of time, the black dwarf itself fades away, which is the fate of all matter as we understand it.

Variable Stars

Stars tend to "twinkle" in the night sky because the narrow beam of light from them gets buffeted about by changes in our atmosphere. But there are stars whose actual brightness varies, or at least whose apparent brightness to us varies. These, not surprisingly, are known as variable stars, and they can be intrinsic variables (whose brightness really does vary for some reason), or extrinsic variables (for instance where a pair of stars eclipse each other).

Representing the intrinsic variables, we have the RR Lyrae variables (lyrids) which are pulsating old giant stars with low luminosity, having absolute magnitudes ranging from about 0.3 to 0.6. Mostly, they are Population II stars found in the galactic halo and in globular clusters. When they are found within clusters they might also be known as "cluster variables," but that term can apply to any variable star found in a cluster. The time between maximum and minimum brightness is short, usually less than 24 hours. Their fairly consistent absolute magnitudes make them useful as another "standard candle" for gauging distances. RR Lyrae (the star that this category is named after) is about 900 light-years away from us and varies from magnitude 7.1 to 8.1 every 13 hours 36 minutes. It is the brightest known RR Lyrae variable, although you'll still need at least binoculars to see it.

Other types of intrinsic variables include eruptive variables and cataclysmic or explosive variables. The former varies due to flares or mass ejections erupting from their surface, and the latter undergo some event like a nova. That might be caused by them pulling matter off of a binary partner star and the matter landing explosively on the star. Supernovae would be an extreme form of explosive variable star[lxxv].

Extrinsic variables include eclipsing binary variables, also known as "beta Lyrids", which are usually Type A or B stars (hot young giant stars) that are in close, rapid orbits. That closeness means that there's a lot of gravitational pull from one star on the other, normally causing the stars to be distorted into ellipsoids. As they orbit each other, we see them both at one point and then only one. As a consequence, they exhibit dimming and brightening over periods that are greater than 1 day, and the light curves vary continually and have a secondary minimum at the midpoint. That's because at one point we are seeing mostly the dimmest star of the pair, at the secondary minimum we are only seeing the brightest star, and at other times we get the combined brightness of the two. Beta Lyrae (the star that gave this category its name) varies from magnitude 3.4 to 4.1 every 12 days 22 hours with a secondary minimum of magnitude 3.8. It can be easily observed throughout its cycle by amateur astronomers using the nearby star Gamma Lyrae (in the constellation Lyra) at magnitude 3.25 as a reference for comparison. Binoculars or a small telescope are helpful. Stars can also exhibit variability as a result of their rotation if, for instance, there is a high volume of sunspots on one side of the star.

Supernovae

Globular clusters are dense groupings of stars that orbit together around galaxies, and the Milky Way has around 150 such globular clusters. The stars in these clusters are some of the oldest stars we see, and they look reddish because the stars cool down with age. But occasionally we find a bright bluish star among them, looking like a young visitor in an old people's home. However, these stars are just as old as the others, but they have been rejuvenating themselves, drinking the lifeblood from a companion. These bright blue stars are often found as part of a binary pair, and the sibling star is seen to be a white dwarf. Earlier in its life, the star that is now a white dwarf had grown into a red giant as it swelled up due to excess heat as a result of it starting to fuse helium instead of hydrogen. As it swelled up, its outer layers got close enough to the other star orbiting with it to allow the other star to suck in a lot of that gas and give itself an infusion of hydrogen. So, the vampire star now shines young and bright, and the first star is left as a "living-dead" zombie that is quietly plotting its revenge. As the vampire star ages, it will also start to swell up and, being in close orbit with the dense zombie with its powerful gravity, the zombie can start sucking gas back off the vampire. With white dwarfs, we know that they can only grow to a certain size (about 1.44 times the mass of the Sun) before the mass triggers them to explode, going out in a bang as a type Ia supernova. And our exploding zombie will destroy its vampire companion in the process. Revenge is sweet, even for stars[lxxvi].

That may seem to be incredibly destructive and violent, but that is how the universe grows up. During a star's lifetime it builds up some of the elements that planets and living beings need, and the vampire star steals the hydrogen fuel that the other star is burping out as it ages. That older bloated star wouldn't be using that fuel, but the vampire star can use it to build up more of the heavier elements. When the zombie explodes, it blasts a lot of the elements created by both stars out into the galaxy to make them available for the next generation, and the energy from the blast helps create even heavier elements in the process. Even cosmic vampires and zombies have a purpose and here they are cooperating to make maximum use of resources and then pass their legacy on to future generations.

Cosmic rays, consisting of particles accelerated to close to the speed of light, are believed to mostly come from supernovae. They are also normally associated with very high-energy gamma rays which help determine the source of the cosmic rays because, while cosmic rays are deflected by magnetic fields that they encounter en route to us, gamma rays are not affected that way. One source of cosmic rays has been identified as RS Ophiuchi, about 5,000 light-years away, consisting of a white dwarf and a red giant. Every 15 to 20 years, the white dwarf brightens in a nova explosion, presumably as material from the red giant gets pulled onto it, giving radiation less than expected from a supernova but between 10 and 100 times as strong as that expected from a more regular nova outburst[lxxvii].

White dwarfs can reach the limit of about 1.44 times the mass of the Sun (the limit at which they explode) by other means too, such as colliding with another white dwarf. Whatever way it happens, the inward pressure of gravity heats the star remnant up enough to start carbon fusion, and that becomes a runaway process that blows up the star. Type Ia supernovae are useful to astronomers because the way they explode lets astronomers calculate how far away they are, and therefore how far away the galaxy is that they are in.

But there are other supernova types as well. Astronomers noticed that some supernovae showed indications of hydrogen in the gas expelled in the explosion, while others didn't. The ones that didn't show signs of hydrogen were called Type I and the others were called Type II. But things are never quite that simple in the universe. Some of the most spectacular supernovae come about when a giant star's core collapses after it begins producing iron as a result of the fusion process. Producing iron absorbs energy, whereas fusion processes up to that point in the star's life had produced energy that could push outwards and counteract the inward push of gravity. So now gravity has no competition, and the whole star suddenly collapses in on itself and then rebounds against the dense core that becomes a neutron star as a massive explosion sends material and radiation hurtling out into space. The shock waves move outward, ripping the star apart. Neutrinos create hot bubbles of material that rapidly expand away from the neutron star and provide the energy to keep the shockwaves moving. The discovery of a stable form of titanium, along with chromium, iron, and similar elements in the supernova remnant known as Cassiopeia A gives confirmation of this idea because these types of elements were predicted to form in such neutrino-driven bubbles[lxxviii]. That kind of explosion could disrupt any solar systems nearby, but it also spreads the material needed to create new solar systems and to support life. Supernovae can create most of the elements we know about, except for the very heaviest ones, but supernovae also lay the groundwork for those too.

Not every supernova will leave behind a compact object such as a black hole. When a star weighs between 130 and 200 solar masses, photons in its core become so energetic that they change into electron-antielectron pairs that can't fully combat gravity. That causes the star to become unstable and, after going supernova, leaves nothing behind because the explosion blasts all of the star outwards.

The reason some supernovae do not show hydrogen in their spectra is because they have either used up most of their hydrogen or have blown off their outer hydrogen layers as they heated up and expanded. The universe is full of variety and sometimes things can look similar but have come about by different processes or look different but have had very similar life cycles. Consequently, there are many subdivisions of the Type I and Type II supernovae. But all of the supernovae are needed to give us the elements that we need to exist and live, and the objects that get left behind after a star goes supernova can be neutron stars which can create the very heaviest elements, including gold, when they collide.

Another type of supernova is called an electron-capture supernova and they occur only with stars between 8 and 10 solar masses. As they age, this size of star will start to build up magnesium and neon atoms in their core that begin to capture the free electrons around them. That causes the outward pressure within the star to decrease and the core collapses to become a neutron star, and the outer regions accelerate down onto it before being blown back out as a supernova. Such a supernova was observed in March 2018 and has been named SN 2018zd[lxxix].

We are getting a fairly clear picture of what happens with supernovae, but they can still spring surprises that end up providing us with more information. Supernova AT2018cow (which came to be known as "the Cow") erupted in 2018 from a spot about 195 million light-years away and reached its peak brightness in a matter of days rather than weeks and was between 10 and 100 times brighter than the average supernova. Happily, it was possible to observe it in different wavelengths, including X-rays, and that led to the conclusion that it was likely to have been a massive star that only partially blew up, leaving behind a small black hole. Actually, it might have left a neutron star, but a black hole fits the models better[lxxx].

A very intriguing merger-triggered core-collapse super-nova was observed in a galaxy about 500 million light-years away. In 2014, a blast of gamma rays was recorded coming from that region, and then around 2018, another major blast of radiation was noted. It is believed that a massive star had been orbiting a neutron star or black hole, and its orbit was decaying. In 2014, the gravity of the compact object disrupted the giant star's core and then, a few years later, the entire star exploded as it tried to swallow the compact object[lxxxi].

The supernova known as SN 1987A marked the death of a massive star 160 thousand light-years away in the Large Magellanic Cloud, and it was close enough to let us observe the small changes that occurred as the supernova evolved. It has now been observed in every type of light for over three decades. On February 23, 1987, it lit up like a brilliant new star in the southern sky. Comparing images from before and after the explosion, the detonated star was identified as a supergiant (Sanduleak -69°202) that had the mass of several suns and had been burning blue. Once it started developing iron in its core, the core collapsed, and the outer layers fell in on it. As the core reached the density of an atomic nucleus, it stiffened, generating a shock wave that then blew off the gaseous outer layers. In 2019, a neutron star was identified as the remains of the star's original core. The fastest gasses were moving away from that core remnant at about a tenth of the speed of light.

For a couple of years, the supernova glowed from radioactive elements created in the explosion, which then decayed and emitted energy that heated the gas. By 1990, the radioactivity levels had dropped and were no longer powering the expanding remnants, so they cooled and dimmed. By 1997, those remnants started brightening again because the high-speed gaseous shock front that had rushed away from the explosion site had slammed into surrounding interstellar gas and heated knots of the interstellar material which released light to get rid of some of the excess energy. Images showed those illuminated patches looking like glowing pearls on a necklace. The Chandra X-ray Observatory was launched in July 1999, and it gave us an X-ray view of the supernova. Images have been taken at about 6-month intervals since then, and the hot gas glowing in the ring is measured at around one million kelvin (1.8 million degrees F). That ring of material had been blown out about 20 thousand years before the blast, maybe because the star was rotating too fast to hold it, or two massive stars merged back then and had flung off the gas.

The glowing of the ring was observed for about twenty years but by 2020 it was growing fainter. New clumps of energized matter were appearing as the blast wave moved out beyond that ring, indicating more matter that the star lost even earlier. There is expected to be a secondary shockwave, which is a bounce back towards the neutron star from the collision with the ring.

The ring remains hot, but the rest of the remnant is frigid, the heat having radiated away. High-energy X-ray radiation shows radioactive titanium-44 scattered across the remnant, revealing how turbulent the motions inside the progenitor star were during the blast, and also showing that the blast was lopsided. Astronomers have mapped the hydrogen, iron, silicon, calcium, magnesium and oxygen in the expanding remnant, which also show the asymmetric nature of the blast[lxxxii].

These expanding remnants seem to be very important in the birth of stars. Our solar system exists in the void within what is known as the Local Bubble. That is a shell of molecular gas from the supernova explosions of around fifteen giant stars that occurred during a period ranging from 14 to 2 billion years ago, and denser regions of that bubble are forming stellar-nursery areas[lxxxiii]. The star-forming regions seem to exist where the expanding bubble is impacting the surrounding gas and dust in the galaxy and compressing it[lxxxiv].

Galaxies

Stars are not normally found on their own, but in collections that are called galaxies. The word "galaxy" comes from the Greek word for "milky" because the collection of stars in our galaxy makes it look as though someone has spilt milk across the sky. As a consequence of that, our galaxy is known as the Milky Way and it is believed to contain about 300 thousand million stars, and for a long-time people thought it was the complete universe. Then it was realized that some of the things people were seeing were outside of the Milky Way and were other galaxies or "island universes" as they were known for a time.

When Edwin Hubble arrived at the Mount Wilson observatory, people were unclear about what the spiral "nebulae" were, but while using the 100-inch Hooker Telescope he looked at the Andromeda nebula (M31) and found a suspected nova. He soon realized it was a Cepheid variable, and its faintness indicated that it might be about a million light-year away, which was three times further away than the then estimated edge of the universe. We now know that the Andromeda galaxy is 2.5 million light-years away. The spiral nebulae became reclassified as galaxies[lxxxv].

Galaxies are massive aggregations of stars, gas and dust and they vary tremendously in size, mass, and form. It is hard to strictly define the size of a galaxy because the stars thin out as you get to the "edge", giving no clear cut-off point. There is also a halo of old stars and dark matter surrounding the main body of the galaxy. So, the Milky Way's disk may appear to be around 130,000 light-years across, but its dark matter halo might be about two million light-years in diameter.

The Hubble Deep Field image looked at a single dark area of the sky in Ursa Major for over two weeks and that small area showed more galaxies (several thousand) than we'd seen before. Extrapolating over the entire sky gives maybe 100 to 200 billion galaxies or more. In 2009, during the final Hubble servicing mission, the Wide Field Camera 3 (WFC3) was installed, taking very deep pictures over relatively large areas of the sky at infrared wavelengths. Observing galaxies over many wavelengths allows us to gauge their mass and work out if that distribution changes over time, and examining their spectrum gives us an estimate of their distance. The further back in time we look, the more low-mass galaxies we see, with maybe 2 trillion galaxies in the early observable universe[lxxxvi]. They were drawn together by gravity and merged over time to give us the large galaxies that we see today. When the universe was just a billion years old, there were around ten times as many galaxies as we see today, and then the galaxies evolved and grew through mergers.

The Hubble images show us galaxies colliding, and we can see evidence of our Milky Way having absorbed smaller galaxies in the past. We also know that we are heading for a collision and merger with (or takeover by) the Andromeda Galaxy. We see disk-like shapes forming in relation to galaxies and solar systems and, in both cases, it has to do with the preservation of angular momentum, and planets, stars and galaxies all spin due to conservation of angular momentum. When an atom, rock, or whatever falls into a gravitational well, unless it falls directly in, it produces a tiny torque which becomes accentuated as the clumps of matter collapse (the normal comparison used for illustrating angular momentum is that of a spinning figure skater pulling her arms in and spinning faster as a result). The Big Bang gave linear motion, gravity turned it into orbital motion[lxxxvii], and objects have to spin or fall directly in. That spin flattens the incoming matter out into a shape more like a disk than a sphere. Mergers of galaxies mess up the nice disk shape for a while, but those same forces that formed the disk in the first place will still be operating to pull the combined mass of stars back into order over time. The space between galaxies that are not gravitationally bound together is continually growing, but galaxies themselves are only growing by mergers.

As the gas cloud condenses to form a galaxy, the densest part will be in the middle, and this is where we expect to find a massive or supermassive black hole. Even small galaxies can have large black holes and they can help or hinder star growth in the galaxy. The black hole can provide the gravity needed to pull more gas into the galaxy and encourage the star-formation, but when the black hole starts feasting on the gas, it can radiate energy that heats up the other gas in the galaxy and stops stars forming. So, the supermassive black hole controls the growth and aging of the galaxy. In larger galaxies they may slow down star growth nearer the center by overheating the gas but encourage it further out where the gas from the black hole's jets ends up. Thirteen supermassive black holes were recently found in dwarf galaxies that were over a hundred times less massive than the Milky Way, half of the black holes being in the center, and the others on the outskirts of the host galaxy[lxxxviii]. It is assumed that the supermassive black holes would have formed in the center of the galaxy, but the cosmos is dynamic, and things can change.

Eight billion light-years away, galaxy 3C186 has a bright galactic object indicating an active supermassive black hole. However, it's about 35 thousand light-years from the center of the galaxy, suggesting it is in the process of being kicked out. Even if the black hole leaves the galaxy completely, the galaxy will continue as normal because the mass of the black hole is negligible in comparison with that of the combined stars, gas and dark matter. Sagittarius A* (Sagittarius A-star - the Milky Way's supermassive black hole) is only one-millionth of the total mass of the galaxy and dominates only the stars and gas in a very small central volume. The one in 3C186 is probably the result of a merger of galaxies and the combining supermassive black holes got a kick from the emitted gravitational waves which then gave it a velocity of several thousand miles per second[lxxxix].

When galaxies collide and merge, their black holes will also ultimately collide and combine as well, but the recoil from the collision can kick the combined black hole away from the center, at speeds up to 310 miles/second (500 km/sec) or more. It could get ejected or might end up at the galaxy's edge and drift back to the center over millions or billions of years. Nine off-center black holes were found in one survey[xc]. Finding two supermassive black holes in one galaxy isn't unusual, and now NGC 6240, about 300 million light-years away, has been shown to have three supermassive black holes, each weighing more than 90 million Suns (the black hole at the center of the Milky Way is about 4 million solar masses), and all three are in a 3,000-light-year wide region (less than 1% of the galaxy's overall size)[xci].

If you'd like to see two supermassive black holes collide, you can point your telescope at galaxy NGC 7727 which is about 89 million light-years away. It has a supermassive black hole with about 154 million solar masses at its core, and another with about 89 million solar masses approaching it. That second one is only about 1,600 light-years from the larger one and is presumably from a small galaxy that got swallowed by NGC 7727. Hang around for about 250 million years and you'll catch the big event[xcii].

Young galaxies are full of gas and dust, fueling star formation, and supermassive black holes in the center grow by eating gas and dust. As the supermassive black hole feeds on nearby material, it belches out energy that heats the remaining gas in the galaxy, preventing it from cooling and clumping to form stars. Galaxy CQ4479, 5.25 billion light-years away has an active black hole and a healthy rate of star formation (making 100 solar masses' worth of stars per year). This will only occur for a limited time period, and this seems to be a last major burst of star formation as the galaxy moves from being in an active stage to entering a quiescent era because of the heating of the gas[xciii]. Smaller galaxies that are more densely packed with stars tend to have larger supermassive black holes, cutting off star formation sooner, while larger diffuse galaxies can evolve more before their supermassive black hole gets big enough to affect star formation[xciv].

The Hubble Deep Field images show that galaxies appeared very soon after the first stars emerged on the scene. The belief is that smallish (on an astronomical scale) dense areas of gas and dark matter became star-breeding areas and developed into small galaxies, and these got drawn to one another and grew into larger galaxies[xcv]. Astronomers have seen galaxies from as far back as 13 billion years ago, and they probably go back further than that. One galaxy, observed as it was 1.5 billion years after the Big Bang, appears to have already finished forming stars at its center[xcvi]. One survey showed that early galaxies apparently underwent a growth spurt between 1 billion and 1.5 billion years after the Big Bang (when the universe was only 10% of its current age). Out of 118 galaxies studied, 20% were very dusty showing they had evolved faster than expected, but how that happened is still unclear[xcvii].

Active galactic nuclei (AGNs) are where the central black holes have beams of energy blasting out perpendicular to the plane of the galaxy. This seems to sometimes create bubbles of less-dense space above and below the galaxy, with the result that there's less gas for satellite galaxies to have to push through. In denser regions further away, the star-forming gas can get stripped out from satellite galaxies, resulting in them no longer being able to form new stars[xcviii].

Two galaxies, NGC 1052-DF2 (DF2 for short) and NGC 1052-DF4 (DF4), both hold far fewer stars than the average galaxy, to the point where you can see distant galaxies through them. They also have large bright globular clusters, and the speed of these clusters indicate that these galaxies have very little dark matter. But that depends on the distance to the galaxies. One group claimed that DF2 was about 65 million (later revised down to 61 million) light-years away and that DF4 was 65 million (+/-9 million), but another group says the figures are more like 42 million and 46 million. Distance estimates utilize various methods including stars at the tip of the red giant branch, which all have basically the same intrinsic brightness. If the lower distance values are correct, the movement of the globular clusters would not be unusual, and the galaxies would have around the expected amount of dark matter[xcix].

A later study of 324 dwarf galaxies found that nineteen of them seemed to have no appreciable amounts of dark matter. It was thought that dark matter was needed to drive galaxy formation, so this is puzzling. If they really do have very little dark matter, it might seem that some very dense gas would be needed to bring these galaxies into existence[c]. Another ultra-diffuse galaxy, similar to DF2 and DF4, is known as AGC 114905, and a careful study of the orbital speed of its neutral hydrogen gas appears to confirm that it has little, if any, dark matter in it[ci]. Maybe galaxies can sometimes form in the absence of dark matter, but they only manage to achieve this kind of "see through" status. We do need to throw more light on dark matter.

Galactic collisions are how galaxies usually grow but the stars are so far apart that smash-ups between stars are unlikely. Turbulence caused by the galaxies' initial pass through each other can trigger gas in the two galaxies to collapse, giving numerous regions where a new generation of stars are formed in a starburst[cii], but galaxy collisions can sometimes be less helpful for star formation. A distant galaxy has been seen to be ejecting nearly half of its available star-forming gas, roughly 10,000 Sun's worth of gas per year, probably as a result of a recent collision with another galaxy[ciii]. Gravity will twist and bend the galaxies out of shape and rearrange their contents until they settle into a final form – a process taking hundreds of millions to billions of years[civ], and sometimes producing very unusual effects. The Black Eye galaxy, NGC 4826, has a gas cloud in the inner region spinning one way, while the gas in the outskirts revolves the opposite direction. The colliding gas at the junction is giving birth to new stars. This odd motion is believed to be the result of a recent galaxy merger[cv].

As we see all over the place, the universe seldom has just one way of doing things. Rubin's Galaxy, UGC 2885, 232 million light-years away and the largest known galaxy "near" us, spans twice the width of the Milky Way and has ten times as many stars. However, it is isolated in space and shows no signs of having grown by smashing into and consuming other galaxies. It may have simply grown by pulling in gas from intergalactic space[cvi].

Edwin Hubble classified galaxies into spiral galaxies, barred spiral galaxies, lenticular (lens-shaped), and irregular (clouds of stars and gas lacking an organized shape). Later, peculiar galaxies were added to the list, and they are ones that seem to have been wracked with explosive or disruptive events. We also have dwarf spheroidal galaxies that were numerous in the early universe. Isolated dwarf spheroidal galaxies consist exclusively of old stars, with star formation having stopped about 10 billion years ago. Donatiello I is an example, and it is suspected of being the most isolated dwarf galaxy in the Local Group.

Now let us look at the three main types of galaxies – Disk, Elliptical and Irregular.

Disk Galaxies

Disk galaxies are comprised predominantly of a thin disk of stars and gas surrounding a central bulge, and they are classified into two sub-types, lenticular and spiral. Disk galaxies tend to fall towards the blue portion of the spectrum, thanks to the light of young hot stars.

While we think of a disk galaxy as just a thin bright disk

and central bulge, there is actually a sphere of stars at least the same diameter as the disk and probably a lot larger. The central bulge is just the densest part of this much larger galactic "halo." Halo stars orbit the galaxy's core in all directions, with orbits from circular to highly elliptical. The stars in the halo are very old, mostly as old as the galaxy itself. There is little or no star formation taking place within the halo because the gas that makes new stars has been pulled into the galactic disk. Thus, halo stars are old, metal-poor and of low mass (mostly red dwarfs and low mass Sun-like stars), plus remnants of high mass stars that are now white dwarfs, neutron stars, or stellar-mass black holes. This population of old, metal-poor stars is what we call Population II.

Within the galactic halo there are very dense clusters of stars known as Globular Clusters, which contain from tens of thousands to millions of stars. They can vary from about 60 to 500 light-years in diameter, but they only account for about 1% of the halo stars. The stars in a globular cluster orbit the center of the cluster in all directions and varying orbital shapes, just like the regular halo stars do. The star density within these clusters is much greater than in the spiral disk, with the result that stellar collisions sometimes occur. Such collisions can be violent and explosive or have the stars merging gently and becoming large hot blue stars, giving more of the "blue straggler" stars that look like bright young stars among the older stars of the cluster. There may be "intermediate mass" black holes in the centers of at least some globular clusters. Globular cluster

NGC 6397 might have many stellar-mass black holes rather than a single intermediate-mass black hole[cvii].

In fact, it seems that globular clusters can have quite a lot of mostly stellar-sized black holes. Palomar 5 is a globular cluster that is being torn apart by the Milky Way, and astronomers have been able to study it and count the stars and black holes, and they found far more black holes than expected in relation to the number of stars. The black holes came from collapsed giant stars that had reached the end of their lives, but the stretching of the globular cluster by the Milky Way led to more interactions between the stars and the black holes, and many of the stars got tossed out of the cluster. By the time the cluster is completely torn apart (in about a billion years' time) it is expected that there will not be much more than black holes left in it[cviii].

Almost all galaxies appear to have globular clusters, from just a few to tens of thousands. The Milky Way has about 150 of them. Astronomers first figured out that the galaxy center was near Sagittarius because that's the center of globular cluster distribution in the sky. Since that is the case, we mostly see globular clusters around the summer months, when Sagittarius is high in the night sky.

Stars in a globular cluster are relatively tightly packed together, but we also have open clusters, where the stars are more spread out. The Pleiades is one example of an open cluster and, in fact, all but two of the stars in the Big Dipper are part of an open cluster known as the Moving Ursa Major Group. It might be tempting to think that open clusters result from globular clusters that have drifted apart over time, but that is far from the case. While globular clusters are some of the oldest stars found in the galaxy and exist mainly in the galaxy's halo, open clusters are the result of recent star forming regions in the disk of the galaxy, and the stars are not solidly bound to one another by gravity, so they go their own way over time. Our Sun would have started out as part of an open cluster. About 1,200 open clusters have been identified in the Milky Way[cix].

Stars near the center of a galaxy need to watch out. Being so close to a supermassive black hole means that they orbit it very fast, and it has been calculated that a collision between two such speeding stars (often traveling at a good proportion of the speed of light) could result in an explosion that would outshine any supernova. That might then be followed by an even larger flare of energy as the remains of the collision get pulled into the black hole. Happily, space is big, so such collisions are rare[cx]. If a star runs into the galaxy's central black hole, it goes out in a tidal disruption event where the material gets drawn into the hot disk of gas around the black hole before being consumed or expelled and thrown away[cxi].

For the first sub-type of disk galaxies, we have lenticular or "lens-like" galaxies that form a disk that is uniform with no spiral structure showing. The second sub-type comprises the spiral galaxies that form a disk containing bright spiral-shaped arms, and this type of galaxy is further subdivided into barred and non-barred spirals. Barred spiral galaxies have a straight, bright "bar" that extends from the center, usually with a major spiral arm originating at each end. Non-barred spiral galaxies only have the bright central spherical region, and the spiral arms come directly off of it. Spiral galaxies can also be classified by the number of arms and how tightly wound they are.

Somewhere between a third and a half of all spiral galaxies have some form of central bar structure, and our Milky Way is one of them. The Milky Way's barred structure spans about 7,000 light-years. Such bars are caused by gravitational instabilities in the motion of material in the galaxy's bulge, and they are simply dense regions where material is closer together, like the spiral arms, and they can change over time[cxii].

Stars that are within the disk normally orbit the galactic center in fairly circular paths. It might appear that most of the stars in a spiral galaxy are in the arms, but the density of stars in a spiral disk is fairly uniform, with the spacing of stars between the spiral arms being very similar to those within them. The same stars don't remain in a particular arm – the arms are density (pressure) waves that spiral from the edge of the disk to the center and back out again, and stars and gas bunch up in the wave crests.

Any particularly dense pockets within these pressure waves can condense into new stars, and so we get star-forming nebulae that are also known as "H II Regions". A small proportion of these new stars are supermassive and thus super luminous. These super-bright, blue-white stars are near the pressure wave of the spiral arms where they formed, making the spirals stand out. Such super-bright stars have short lives and explode in supernovae before they travel very far from the spiral arms where they formed.

The formation of spiral arms in galaxies can involve a combination of processes. In a spiral galaxy, everything orbits at the same speed, so stars and gas near the center complete one orbit faster than those further out (differential rotation). That will pull on the gas in the galaxy in a twisting fashion, but that, by itself, would lead to many tightly bound spiral arms, whereas we usually see only 2 to 4 main arms. Planetary rings and protoplanetary disks can also have density waves and spiral structures. In planetary rings, small moons can cause waves, and in protoplanetary disks the density waves are driven by planets forming[cxiii]. The spiral structure becomes a visual sign of on-going new star formation, and such star formation is unique to the disks of spiral galaxies.

Spiral galaxy disks therefore have stars of many ages, from very old to very young. Similarly, they have a full range of star masses, from low mass to supermassive. The oldest stars formed from hydrogen and helium only and are "metal poor", whereas younger stars form from gas and dust that includes heavier elements formed in prior stars, and these are "metal rich". This distinctive mix of star types is characteristic of spiral galaxy disks and is what we call Population I.

Elegant spiral galaxies are believed to have arisen by capturing smaller galaxies, some possibly only composed of dark matter. An assortment of large-scale tidal structures has been seen around nearby massive galaxies, and we have also observed interactions between the Magellanic Clouds and other Milky Way satellite galaxies. Computer models suggest that the halos of massive galaxies should contain tidal debris streams, wrapping around the host galaxy and roughly tracing the orbit of the satellite galaxy that they originated from. That is reminiscent of the debris from comets that give us meteor showers. One of these galactic debris streams is the Sagittarius Stream surrounding the Milky Way. It seems that the absorption of smaller satellite galaxies plays a significant role in building up stellar halos around galaxies and can trigger starbursts as well. A dense, shell-like area of stars was seen in the outskirts of the Small Magellanic Cloud composed of young stars probably triggered by an interaction with the Large Magellanic Cloud about 150 million years ago[cxiv].

Early efforts to simulate colliding galaxies indicated that close encounters could generate spiral arms and simulating mergers of two spiral galaxies showed the likelihood of forming an elliptical galaxy. Interestingly, elliptical galaxies usually huddle in groups and clusters, while spiral galaxies prefer to have more room around them.

Elliptical Galaxies

Elliptical galaxies are big "fuzz-balls" of stars, and they vary from spherical to elongated ellipsoids (i.e., they can show "eccentricity"). They tend to be in the red portion of the spectrum, indicating that they contain older stars and that there are few, if any, new star-forming regions.

It is believed that spiral galaxies probably formed first, and elliptical galaxies are the result of collisions between galaxies that destroyed their structure. It is probable that the merger of the Milky Way galaxy and Andromeda galaxy in about five billion years will result in a large elliptical galaxy, with little in the way of new star formation after the burst of new stars resulting from the colliding gas of the galaxies. But as the Milky Way and Andromeda approach and pass through each other, gravity will draw them into very irregular shapes before finally pulling them into an elliptical pattern.

Irregular Galaxies

The term "irregular galaxies" is a catch-all phrase that covers all other galaxies. They have no regular structure and can best be described as a "mess". They are believed to have resulted from galaxy collisions or near-misses and were not formed the way that we see them today. Over time, they may pull themselves into something like an elliptical galaxy.

The Milky Way

The bright band of the Milky Way that we see in the sky is from the glow of millions of stars seen as we look out through the plane of the galaxy's spiral disk. Gas and dust block visible light from more distant stars, with the result that we can only see about 8,000 light-years on average into the spiral disk using visible light, and that has made it awkward trying to establish what the galaxy looks like overall.

Our galaxy, the Milky Way, was probably built up from dozens or more protogalaxies, and it has several major parts, with the most recognizable being the disk where most of the stars, gas, and dust reside. There are something like 300 billion stars in this 130,000 light-year diameter disk, and our Sun is one of those stars. 25 billion of those stars could be very much like our Sun and maybe 20% of those could have an Earth-sized planet orbiting them. The much more numerous and much longer-lived red dwarf stars appear to commonly have Earth-sized planets in their "habitable zones" as well.

It is believed that our Milky Way galaxy started growing from the merging of dwarf galaxies about 13.6 billion years ago (around January 8 on our "cosmic year" scale) and it has continued to swallow up smaller galaxies. A study of fifteen extreme-velocity stars showed that eight of them have a different ratio of elements from the majority of stars in the Milky Way. That implies that they most likely origi-nated in other galaxies that got absorbed and torn apart by the Milky Way[cxv]. Much of the gas settled into the center of the region where the galaxy was forming, increasing the spin of the gas due to preservation of angular momentum and that spin flattening the bulk of the mass out into a re-volving disk by around nine billion years ago (May 7 of our "cosmic year"). Stars, including our Sun, continued to form, and the galaxy pulled in more star-forming gas from the intergalactic region whenever it had the opportunity.

The Milky Way is known to have already consumed dozens of smaller neighbors, and we can see trails of stars tracing giant arcs around our galactic disk. The Sagittarius dwarf galaxy is known to have passed through the Milky Way at least three times, and it has now been stretched out as a result of those encounters. But it has also shaped the

Milky Way, maybe being at least partially responsible for the galaxy's arms, and bursts of star formation seem to correspond with its transits through the galaxy[cxvi]. Collisions with such small galaxies are getting rarer as individual galaxies and galaxy groups become, on average, further and further apart.

There are four very long spiral arms that we know of. The two major arms are Scutum-Centaurus (a.k.a. the Centaurus arm) which starts at the near end of the bar to us, and the Perseus arm which starts at the far end of the bar. The two lesser arms are Carina-Sagittarius (a.k.a. the Sagittarius arm), which is parallel to and outside of the Centaurus arm, and Norma-Outer (a.k.a. the Norma arm), which is parallel to and outside of the Perseus arm. The Centaurus arm is totally hidden from our view by the Sagittarius arm on the side closest to the galaxy's core, and by the Perseus arm further out.

The Sagittarius arm has been observed to have a spur jutting out of it at about a 45° angle, consisting of a 3,000-light-year-long collection of stars and gas. Such spurs (also known as feathers) have also been observed in other spiral galaxies, but this is the first one found in our galaxy, although others are expected to exist as well[cxvii].

A filament of gas, at least 16,000 light-years long by 675 light-years wide, containing gas equivalent to 65,000 Suns, has been observed at the edge of the galaxy, about 72,000 light-years out from the galactic center. It has been named Cattail, and there is speculation that it may be part of an otherwise unobserved outer arm of the galaxy[cxviii], or an extension of the Centaurus arm.

At our location, the distance between spiral arms is about 4,000 to 8,000 light-years. The Sun is passing through a minor arm known as the Orion arm, also known (not surprisingly) as the Local arm. It spans between the Sagittarius arm and the Perseus arm, and most of the stars that we see in the night sky are associated with this Orion arm. Our Sun is following the constellation Cygnus as it orbits around the galactic core but is shifting over in the direction of Hercules.

The galaxy's disk is encapsulated by a halo consisting of a relatively small number of stars and large collections of ancient stars called globular clusters, along with a huge envelope of dark matter. The galaxy is currently absorbing gas from the Small and Large Magellanic Clouds, and from a 10,000 light-year-long by 3,000 light-year-wide hydrogen and dark matter cloud known as the Smith Cloud, which is heading towards us at 200,000 mph (320,000 km/hr.) and is expected to start colliding with the Perseus Arm in twenty-seven million years.

It has been suggested that the Milky Way was violently warped by the Large Magellanic Cloud about 700 million years ago, and that that satellite galaxy is still pulling the Milky Way in the direction denoted from our viewpoint by the constellation Pegasus.

At the center of the Milky Way is our supermassive black hole, Sagittarius A*, which has a mass equal to 4 million Suns. It was expected that stars near the galactic center would be relatively old because clouds of gas and dust would be unstable so close to the black hole since they would be disrupted by tidal forces, yet stars near the center seem to be on the young side. I suppose it is possible that they are WIMP-burners, but there may be a more traditional explanation that would emphasize the fact that we don't fully understand the dynamics of galaxies yet. Stars have been observed being born within 1,000 light-years of the center of the Milky Way, so maybe star formation isn't as difficult there as previously imagined[cxix]. It seems that about 80% of the stars in the galaxy's center formed between 8 and 13.5 billion years ago, and that another intense burst of star formation occurred about 1 billion years ago creating many massive stars[cxx].

Star SO-2 (also known simply as S2) is one of the thousands of stars orbiting fairly close to Sagittarius A*, our supermassive black hole. It takes only 16 years to orbit Sagittarius A*, whereas the Sun takes 230 million years to orbit the galaxy[cxxi]. Tracking S2 in its highly elliptical orbit using the 10-meter Keck telescopes shows that it made a close approach to Sagittarius A* in mid-2018. At that time, it was about three times as far from the black hole as Pluto is from Sun, and it was moving at one to two percent of the speed of light. The orbits of S2 and a few similar stars really pinpoint the presence and position of our supermassive black hole.

Vast clouds of X-ray emitting gas, like giant bubbles of hot plasma about 50,000 light-years long, have been observed above and below the Milky Way, which seem to be the result of an active feeding phase of our supermassive black hole, when it was behaving rather like a quasar or blazar. The northern cloud is denser than the southern one, implying that Sagittarius A* was selective in the direction that it spat out its excess food while it was eating[cxxii]. It seems to be largely the strong magnetic fields around our supermassive black hole that makes it such a messy eater, and another indication of strong magnetic fields near the center of the galaxy is the lack of cosmic rays that we observe emanating from that region. Cosmic rays are charged particles and can be deflected by strong magnetic fields, such as those associated with a supermassive black hole.

When Sagittarius A* is feeding, we could expect plenty of cosmic rays to result from the beams of energy that it emits, and the beams would also push gas within the galaxy out and heat up any remaining gas. That would cause star formation to slow down, which helps lifeforms because the radiation from exploding short-lived giant stars can be damaging to life or even strip a planet of its atmosphere. The fact that the gas was driven out would also make Sagittarius A* (or any other supermassive black hole) go quiet, which would let the gas cool and the gravity from the black hole and the rest of the galaxy would start to pull the gas back into the galaxy, allowing star formation to start increasing again. It would then start feeding the supermassive black hole once more, and the cycle begins again.

The huge star-forming clouds of gas visible to us in the galaxy's main disk have been found to form what has been called the Radcliffe Wave, which is almost 9,000 light-years long and 400 light-years wide and creates a snaking line of interconnected star-forming regions, rising above and dipping below the plane of our galaxy. It passes less than 500 light-years from us at its nearest point and connects molecular clouds in Orion, Taurus, Perseus, Cepheus, and

Cygnus. Tendrils of filamentary gas link the various star forming regions[cxxiii]. Those tendrils remind me somewhat of the filaments of individual galaxies and gas that connect the nodes of galaxy clusters forming the Cosmic Web. There might be about one new star on average created each year in the Milky Way currently.

Our Sun is about 25,600 light-years from the galaxy's core, or 40% of the way out from the center, orbiting once every 230 million years, and it has completed about 20 rotations of the galaxy since the solar system formed.

As Voyager 1 moved beyond the heliosphere (out beyond the Kuiper Belt where the Sun's magnetic field and solar wind meet their match in the interstellar environment) it detected a faint persistent murmur coming from the empty space ahead of it, and that hum is believed to be coming from low-level, long-lasting vibrations in the interstellar plasma[cxxiv] of the galaxy.

Quasars

Earlier we mentioned that colliding neutron stars produce enough energy to create the heaviest elements, like gold. However, neutron stars don't collide often enough to account for all the gold and other heavy elements, so there must be some other power source out there as well. If you are looking for power, black holes might be a good place to start.

Quasars (quasi-stellar objects) are the most brilliant objects in the universe, and they have been shown to be supermassive black holes that are feeding on the primordial gas in the early universe. When the supermassive black holes in galaxies try eating a gas cloud, they can make a mess. As a black hole pulls gas and other material towards it, the material will normally be pulled into orbit around the black hole, traveling at tremendous speeds and colliding with other material. This maelstrom is called the accretion disk and some of the material gets pulled into the black hole, but a lot gets spun up towards the poles of the super-massive black hole by the magnetic field generated from the fiercely hot plasma[cxxv]. There it gets spat out as a stream of particles and radiation that can be traveling at a significant fraction of the speed of light, blasting adjacent stars and galaxies with radiation that can be called deadly, but is also transformative. The energy beam from one quasar has been measured as being 1.4 megaparsecs (4.6 million light-years) long. Quasars only remain active while there is plenty of gas near the black hole for it to attempt to feed on, so they may only last a few million years.

By the mid-2010s, quasars dating back one or two billion years after the Big Bang were being found, and ULAS J1120+0641 was the first one discovered beyond a redshift of 7, blazing out from a supermassive black hole that is two billion times the mass of our Sun, and dating back to when the universe was only 770 million years old (January 3 in our "cosmic year"). It seems to be way too massive for that time, so soon after the Big Bang. In June 2020, a quasar named J1007+2115 was seen with 1.5 billion solar masses and a redshift of 7.515 that means it dates from when the universe was only 700 million years old. J1342+0928 (J1342 to its drinking buddies), with a redshift of 7.54, is seen radiating 40 trillion Sun's worth of energy from back when the universe was only 690 million years old. The number of quasars declines rapidly with increasing redshift, so not many more from that early era are expected to be found. There may be only 1 per cubic gigaparsec (parsec = 3.26 billion light-years), so conditions for forming these very early quasars seem to have been uncommon. However, it is likely that being a quasar is part of the lifecycle of any galaxy.

Quasar P172+18, from 780 million years after the Big Bang, is the oldest example of a radio-loud quasar, a galaxy with an actively feeding supermassive black hole with jets emitting intense radio waves[cxxvi].

Our universe looks calm and peaceful as we gaze up at the night sky, but it is dynamic and growing. What looks like a deadly explosion can be just clearing out the old and creating the new and making life possible in the process. Happily, most quasars were active when the universe was still relatively young and there was abundant gas to gorge on. That was presumably before much in the way of life had had time to evolve, because the jets of radiation would not be friendly things to have heading in your direction.

Galaxy Clusters, Super Clusters, and Cosmic Web

Galaxies tend to be grouped into gravitationally bound clusters. When galaxies first formed, the universe was a lot smaller, or at least a lot more compact. Consequently, galaxies tended to form not too far from each other.

The region known as EGS77 consists of three galaxies that we are seeing as they were 680 million years after the Big Bang (late on January 18 in our "cosmic year") when the known universe was only 5% of its current age. Each galaxy is seen to be forming its own bubble of ionized hydrogen about two to three million light-years across, with those areas overlapping and creating a large single region of space that is free of cosmic fog and therefore letting light travel freely[cxxvii].

Large clusters may have hundreds of galaxies and be 30 million light-years across or more. The biggest galaxies outweigh the smallest by factors of a million or more, with the largest galaxies tending to be found in the center of galaxy clusters and they are normally giant ellipticals[cxxviii]. The Milky Way is in a small cluster called "The Local Group" that is about 12 million light-years across, about a hundred times the diameter of the Milky Way and it has only a few large galaxies and about 25 dwarf galaxies.

The primary members of the Local Group of galaxies are the Milky Way, the Andromeda Galaxy (M31), and the Pinwheel Galaxy (M33), but each of these has a cloud of smaller attendant galaxies like the Large and Small Magellanic Clouds, plus many dwarf galaxies. There are large assemblies called clusters and superclusters, our Local Group being part of the Virgo Cluster which has at least 1,500 galaxies, with its center around 54 million light-years away from us. Although the universe is expanding, gravity holds such groups of galaxies together.

The average temperature of galaxy clusters is nearly 4 million degrees F (2.2 million degrees C), three times hotter than it was 8 billion years ago, and that rise in temperature is caused by friction as galaxy clusters pull in gas over time. That is the temperature of the atoms in the gas, but the gas is so sparce, with the atoms so spread out, that you'd still freeze if you were out there (you'd radiate a lot more heat away than you could ever gain from the searingly hot but well-spread-out atoms).

A galaxy cluster can hold hundreds to thousands of galaxies, all bound by gravity, along with tenuous gas glowing at millions of degrees between the galaxies, plus dark matter holding it all together. All of that provides an incredible mass to warp space. Consequently, a galaxy cluster can act like a giant lens that bends, amplifies (by about thirty times) and warps light from distant galaxies and supernovae, thereby producing distorted images of any background object. Models are used to analyze the distorted images that often appear in the form of arcs. One supernova (Supernova Refsdal), that was observed by this kind of effect, first produced 3 images, then 4 as light arrived from a longer path around the galaxy cluster, and the model correctly predicted when a fifth image would appear. About 250 dwarf galaxies were found hiding behind three galaxy clusters. Such images and the associated models also help astronomers to analyze the superclusters[cxxix] themselves. Before JWST started its work, the most distant individual star observed, nicknamed "Earendel", which is 12.9 billion light-years away, was observed by the Hubble Space Telescope thanks to such gravitational lensing[cxxx].

Just as galaxies can collide and combine, so can galaxy clusters. In NGC 6338, two galaxy groups are rushing towards each other at about 4 million mph (6.4 million km/hr.), and the hottest gas (36 million degrees F, 20 million degrees C) is heated by shocks from the collision[cxxxi].

Galaxy clusters are, in turn, grouped into gravitationally bound clusters of galaxy-clusters known as Galaxy Superclusters. Our Local Group is believed to be part of the Virgo Supercluster, which is centered around the Virgo Galaxy Cluster, and we are only about 45 million light-years away from the supercluster's center. That Virgo Supercluster, also known as the Local Supercluster, contains about 100 galaxy groups and clusters and spans 110 million light-years across. It is one of about 10 million superclusters in the known cosmos.

In 2014, the Laniakea Supercluster (Hawaiian for "immense heaven") was identified, containing 100,000 galaxies including the Milky Way and the rest of the Local Group, and forming a tangled knot of filaments stretching across half a billion light-years. The whole group is traveling together through space, although not all are gravitationally bound, so some will go their own way in time. This gigantic supercluster contains a total of around 500 galaxy clusters and groups and it has four major components: the Virgo Supercluster, the Hydra-Centaurus Supercluster, the Pavo-Indus Supercluster, and the Southern Supercluster. Surrounding Laniakea are other galaxy superclusters – the Shapley Supercluster, the Hercules Supercluster, the Coma Supercluster, and the Perseus-Pisces Supercluster, each holding hundreds of galaxy clusters and linked by the fabriclike structure of the cosmic web.

Galaxy superclusters are grouped into large membrane-like structures something like the surfaces in a mass of soap bubbles, with vast expanses of largely empty space between. Those giant voids range from dozens to hundreds of millions of light-years in diameter with galaxies "coating" the surfaces. The Great Wall (identified in the 1980s) is a sheet of galaxies 200 million light-years across and about 16 million light-years thick. The more recently discovered Sloan Great Wall is far bigger, almost 1.5 billion light-years in the long dimension. We are being pulled towards a mammoth structure called the Shapley Supercluster that is 650 million light-years away and contains the greatest concentration of galaxies in our local part of the cosmos. It lies in the direction of the Triangulum Australe and Norma constellations in the southern sky. The Giant Arc is an even larger structure, discovered in 2021. This is a collection of galaxies and clusters spanning 3.3 billion light-years[cxxxii].

Beginning in the 1980s, massive structures called Large Quasar Groups (LQG) were discovered. The Webster Large Quasar Group is a collection of 5 quasars that span a distance of more than 330 million light-years. The imaginatively named Huge LQG contains 73 quasars over a distance of about 4 billion light-years, or let's say it is 1.24 gigaparsecs long x 640 megaparsecs x 370 megaparsecs, or thereabouts.

Trying to work out how the universe formed, scientists have built computer models that started with a universe filled with dark matter, and they ended up with a cosmic web of filaments and clumps called dark matter halos within which galaxies are believed to form. Running the simulation for repeatedly smaller regions (in other words, focusing in to see more detail) showed the filament and clump structure repeating to smaller and smaller scales, with the smallest clumps that were simulated being about Earth-size. Within the clumps, the density is greatest in the centers, so that is where it is most likely that dark matter and its antimatter equivalent would collide and annihilate each other, and if that is happening it should be giving off bursts of gamma-ray light that might be detectable[cxxxiii].

Matter is indeed seen to distribute itself in the filamentary fashion that the models suggest, and we see long streams of galaxies woven together into a cosmic web. For instance, the Perseus-Pisces Supercluster forms a chain of galaxies across 50 degrees of the northern sky, fed by smaller tributary-like filaments containing densely populated groups and clusters of galaxies, and between them are immense voids. The Milky Way is on the outskirts of such a structure, the Laniakea Supercluster mentioned above. It has now been observed that some, and probably all, of the filaments tying the cosmic web together are spinning, so they are not just funneling matter down into the galaxy clusters at the nodes, but the matter is spiraling down[cxxxiv].

The clustering of galaxies started early in the universe's life. Six galaxies have been discovered dating back to just 900 million years after the Big Bang (January 24 in our "cosmic year"), and they form a close grouping with a supermassive black hole in the center (SDSS J1030+0524). It is believed that they fed on gas that got stockpiled in a tangled knot of cosmic web filaments. Galaxies tend to grow where the filaments cross, and streams of gas can flow along those filaments[cxxxv] feeding the growing galaxies. That can be a good and a bad thing. The connectivity can lead to more gas to build new young stars, but if the gas feeds the supermassive black hole at a galaxy's center it can send out beams of superhot plasma that heats up the gas in the galaxy. That will prevent the gas from clumping together, so the black hole shuts down star birth in the galaxy. It seems that the more connected a galaxy is, the faster it ages and turns into one with predominantly red cool stars as star birth comes to an end. Even galaxies can get too much of a good thing.

The European Southern Observatory's Very Large Telescope (VLT) focused for 140 hours on a single patch of sky and obtained an image from around 1 billion years after the Big Bang (January 27 in the "cosmic year") that shows cool hydrogen gas filaments of the cosmic web extending more than 15 million light-years, along with billions of dwarf galaxies giving birth to stars[cxxxvi].

Gravity might be the main driver of this network, but the electromagnetic field seems to play a role in the formation of the cosmic web too. A magnetic field bridging galaxy clusters across 10 million light-years has been observed, with charged particles twisting in spirals as they move through the field.

The galaxies themselves are shaped by the cosmic web. There have been indications for a long time that galaxies seemed to favor particular orientations. Spiral galaxies spin out into a flat disk, but elliptical galaxies have little or no rotation with stars orbiting in seemingly random paths. Even so, elliptical galaxies tend to elongate in one direction more than others, shaping the galaxy to look something like an American football. Giant elliptical galaxies that populate the centers of clusters tend to elongate in the same direction as the host cluster. Clusters of galaxies tend to "point" towards neighboring clusters. Clusters in the Perseus-Pisces Supercluster are seen to be elongated in the same direction as the filament that bridges them.

Some of these elongations may be the result of galaxy mergers occurring along filaments in the cosmic web and imparting the end result with a preferred alignment. It is also possible that gravity's relentless tug will slowly orient galaxies to align with their surroundings, and it's likely that both of these ideas are involved to some extent. In 2014 it was reported that the spin axis of some quasars (which are powered by supermassive black holes) are parallel to each other over distances of billions of light-years, and that they align with the surrounding filamentary structure.

Anyway, enjoy the galaxy clusters while you can. If the increasing dark energy expansion continues for more than ten times the universe's current age it will mean that our galaxy will be left with no other galaxies in sight.

The Cosmological Principle

Our models of the universe are normally based on the idea that, as you pan out to show more of the universe, the smoother and more uniformly spread-out matter is. Of course, we know that locally it clumps into galaxies and galaxy clusters, and there can be massive voids of seemingly empty space between them, but on average any large volume of space covering a reasonably large chunk of the cosmic web will have the same overall density of matter and energy as another similar chunk elsewhere. That is called the cosmological principle, but not everybody accepts that idea. I have already mentioned that cosmic rays seem to arrive more frequently from one side of the universe than another, but that could be caused by a more local source creating or deflecting cosmic rays. Making an assumption like the cosmological principle facilitated the obtaining of solutions to Einstein's equations for General Relativity and led on to the current "standard model" of cosmology, which works very well. However, when we start finding structures such as the Giant Arc, that spans over three billion light-years, questions start to be raised as to how homogenous and isotropic the universe really is[cxxxvii]. Also, two analyses of the galaxies that show up in the Sloan Digital Sky Survey gave results that show parity symmetry being violated across the universe. It is assumed that the symmetry must have been broken early in the life of the universe, and there's a possibility that whatever caused the violation might have also accounted for the fact that there is now more matter than antimatter. That would definitely be a good thing, because it means that stars, galaxies, and us can exist[cxxxviii]. Hopefully, the James Webb Space Telescope and other new tools coming available will be able to give us a bigger picture and help resolve these questions before we move on to new ones.

152

Chapter 7: Stellar Evolution

Alchemical Upgrade 2.5: Citrinitas or Yellowness

There is another step in the alchemical process that is called citrinitas or yellowness. It is related to the solar dawn or awakening. Sometimes this step is simply considered to be part of the albedo stage, so we'll call it step 2.5.

As far as the universe is concerned, this citrinitas step could include the time when our galaxy formed and up to around the time when our Sun and solar system came into existence (about 4.6 billion years ago, or 9 billion years after the Big Bang - the universe is patient). It could also include other developments in the lives of stars and galaxies, such as the development of neutron stars and stellar-mass black holes.

Heavier Elements

The cores of stars have only created elements as heavy as iron, and then some heavier elements formed utilizing the power of supernovae. So, what about the heaviest ones, including the gold that the alchemists were always interested in? Be patient and we'll get there, although we've dropped some very broad hints already.

The average body has 7 octillion atoms (7×10^{27} or 7 billion billion billion), most of which were created in the early stages of the universe or in the bellies of exploding stars and the like. You have at least small traces of 60 chemical elements, with oxygen being the most abundant (it's one of the constituents in the water that makes up most of your body), carbon is second, then comes hydrogen (another ingredient of water), nitrogen, plus heavier elements including calcium, phosphorus, potassium, sulfur, sodium, chlorine, and magnesium, and some naturally occurring radioactive elements.

The 118 known elements, organized by their properties in the periodic table, show a wide range of characteristics. The first 94 of those elements are known to occur naturally, and the other 24 have been synthesized in labs and nuclear reactors but have not been seen in nature yet. Stars are nuclear fusion reactors, fusing lighter elements into heavier ones, creating atoms up to iron and nickel. From dying low-mass stars we get about two dozen elements including carbon, nitrogen, strontium, and tin. From supernovae we get another couple of dozen elements including oxygen, potassium, sodium, arsenic, and aluminum. A small number of elements can be created by exploding white dwarf stars, including titanium, vanadium, chromium, manganese, iron, and nickel. From cosmic ray fission, when energetic particles impact Earth's atmosphere and surface, we get boron and beryllium. Finally, from colliding/merging neutron stars we get another couple of dozen elements, including iodine, xenon, cesium, platinum, and gold. We are all part of the universe's recycling program involving the birth, life and death of stars[cxxxix].

The second generation of stars incorporated some of the elements created by the initial short-lived stars, but those new stars were mostly smaller, often similar to our Sun. The reason they were generally smaller than the first stars is because the gas from which they formed had now been spread out more and was not so readily available as a building material for new stars.

When a star around the size of our Sun ages, it starts to run out of hydrogen and then begins to fuse helium in its core. That causes it to heat up, which makes it swell into a red giant, at which time it can blast off the outer layers of gas to form what is known as a planetary nebula. That has nothing to do with planets, but being roundish and glowing, they looked something like planets until they were observed in more detail. When that detail was seen, it was thought that the nebulae might be showing the birth of new solar systems, but now we know that they are more like memorials to the death of one. At the center of the planetary nebula is the core of the old star which, by that time, is about the size of our Earth and is known as a white dwarf. The material in a white dwarf is compressed so much that a tablespoonful of the material would weigh something like ten tons (9 Mg.).

When these Sun-like stars age and puff up into red giants, they can form larger atoms, such as copper, by allowing existing atoms to pick up neutrons that other atoms spit out. Those neutrons then emit an electron and decay into protons, changing the atom from one element to another. This s-process (slow neutron-capture process) builds up heavier elements as an atom absorbs a stray neutron which then decays into a proton in the atom's nucleus. However, as the name implies, the process is very slow, with a nucleus maybe capturing one neutron per month, so the heaviest atoms are not likely to be built up that way.

The r-process (rapid neutron-capture process) is where atoms are bombarded with neutrons, rapidly growing in size, and then decaying through radiation into smaller but still large/heavy elements. This process requires an immense amount of energy, so it needs to be something as powerful as a core-collapse supernova caused by the death of a massive star. This also seems to be the process by which really heavy elements like gold are made, but supernovae don't appear to be able to provide quite enough oomph to win the gold.

The supernovae caused by exploding white dwarfs are different, with those consisting of the explosive ignition of carbon and oxygen that does not provide a rich source of neutrons, so those type Ia supernovae are only likely to give us silicon, sulfur, iron and nickel[cxl].

By methods such as the s-process and by the supernova explosions that provide even more bounteous energy, heavier atoms are formed. However, even supernova explosions may not be powerful enough to form the very heaviest elements like gold. For that we need those neutron stars that we keep talking about.

Colliding neutron stars create an explosion called a kilonova, and that has the power to get the r process producing gold. In a kilonova, it was calculated that the initial flare would be blue as lighter elements were formed, then become red as heavier elements were generated. In 2017, LIGO spotted a signal from two colliding neutron stars, and optical telescopes focused on the source and confirmed the expected light curve. But there don't seem to be enough such kilonovae to explain the quantity of the heaviest elements that we find. It is thought that some of the early massive stars that collapsed down to a black hole might have caused explosions fierce enough to initiate the stronger levels of the r-process, and that seems to be indicated by the fact that some very old stars are found to contain traces of heavy elements like gold. But even then, only about 20% of the gold, etc., can be accounted for from kilonovae and these collapsars[cxli]. However, we have seen

that there are other explosive energy sources, such as quasars, that might be well capable of providing the kinds of
energy needed.

Chapter 8: Neutron Stars and Black Holes

Neutron Stars

If a dying star starts out with a mass around ten to twenty-five times that of the Sun at the time that it goes supernova, instead of collapsing into a white dwarf as a lot of the matter gets blasted out (forming some amazing-looking nebulae), gravity compresses the atoms in the core remnant so much that the particles making up the atoms are pushed together until you get a neutron star. The electron degeneracy pressure that holds up white dwarfs has been overcome and the electrons fuse with the protons to become neutrons. The whole star becomes like one massive atomic nucleus held up by neutron degeneracy pressure. A teaspoonful of material from a neutron star would weigh around 10 million tons (9 million Mg.).

A neutron star can be spinning about 100 times a second on its axis. A lot of the mass gets blown away in the supernova, so the neutron star might end up with a mass that is about 1.5 or 2 times the mass that our Sun has now. They might be smaller than 10 miles (say 10 – 15 km) across, and their gravity is so strong that it pulls the star into an almost smooth sphere, with any surface deformation expected to be less than a millimeter high. We may have a billion neutron stars in the Milky Way, although only about 3,000 have been identified so far.

It is believed that neutron stars have a thin atmosphere of hydrogen and helium over a crust, which is probably less than an inch thick and made up of loose electrons and atomic nuclei. Below that is a region where the electrons are being pushed closer to the atomic nuclei and some are combining with the protons in the lighter nuclei to form neutrons, and the deeper you go, the more neutrons you'll find. In the center of the neutron star, the gravitational pressure seems to be too high for even neutrons to survive, so there may be a core of quarks supporting the bulk of the neutron star above it.

It has also been suggested that neutron stars at a specific density would allow the electromagnetic force and the weak nuclear force to recombine into the electroweak force. That force can turn quarks into electrons and neutrinos, which would make the star lose a lot of mass, and make it give off massive amounts of energy, stopping the star's collapse in the process.

Consisting largely of neutrons which, as their name implies, are magnetically neutral, you might expect that neutron stars have little, if any, magnetic field. That is far from the case, and their magnetic fields are much greater than any such field observed elsewhere in the universe, although exactly how that field is generated is not fully understood[cxlii].

Looking out into the night sky, the stars that we see are often pairs of stars orbiting each other, or there may be three, four, or more stars orbiting together. Having a single star like our Sun is more the exception than the rule. Similarly, we often find pairs of neutron stars orbiting each other, and they can be orbiting very fast and very close together. Two have been seen orbiting each other at around 700,000 mph (1,126,000 km/hr.). However, those orbits decay over time and the neutron stars collide, releasing more energy than the Sun generates during its whole lifetime. The particles in each neutron star are crushed together so densely that the star is basically one massive atomic nucleus and smashing two of those together gives more than enough energy to create gold, uranium, iridium, lead, and all the other heavy elements that we end up using here on Earth.

Neutron stars come in a variety of types that may indicate different stages in a neutron star's life or might show that they have different environments, such as having a companion star to help speed up the neutron star's rotation.

Neutron stars may find themselves as part of a binary pair where the partner is a regular star. Having such a strong gravitational field, the neutron star can pull gas from the regular star, and the hydrogen and helium will then burn on the surface of the neutron star as a burst of heat and energy. As the matter builds up, it will include a growing volume of carbon about 300 feet (100 meters) below the surface, and every few years it will have gathered enough to cause that to burn and produce a superburst that is about 100 times as powerful as the regular bursts. Wait about a thousand years, and sufficient oxygen will have accumulated about 1600 feet (500 meters) below the surface so that suddenly it will burn to give a hyperburst that is a further 100 times as energetic as a superburst and push temperatures up to around 720 million degrees F (400 million degrees C). Such a hyperburst is thought to have been the cause of the neutron star MAXI J0556-332 being so hot[cxliii]. Neutron stars may be called dead stellar remnants, but they won't rest quietly.

Pulsars

Pulsars were the first type of neutron stars to be discovered, with their regular repeating signal looking as if it might be something artificially generated - that is what drew people's attention to them.

Neutron stars have powerful magnetic fields that are

maybe a quadrillion times that of Earth's field, and which have some of them sending powerful focused beams of radio waves in our direction at regular intervals. Those signals have earned them the name of pulsars. The energy beam is controlled by the neutron star's magnetic field which, like Earth's, is probably not strictly aligned with its axis of rotation. So, the focused beam will not be pointing in the same direction constantly. As a consequence, if Earth lies in the beam's path, we will see a quick blip as the beam hits us on each rotation. Of the billion or so neutron stars expected to exist in the Milky Way, only a million or so are expected to exist as pulsars. We have discovered around 2,300 so far. So, it is expected that the pulsars make up about 0.1% of all neutron stars, yet over 75% of neutron stars that we have identified currently are pulsars. That is because pulsars advertise themselves so well.

Pulsars are spinning very rapidly, and their beams of light and radiation shoot out into space from hot spots which were expected to be at each of the pulsar's magnetic poles. Pulsar J0030+0451 is 1,100 light-years away, weighing in at about 1.35 solar masses but it is only about 16 miles (20 km) wide. When mapped, it appeared to have no hotspot at its north pole, but 2 or 3 near the south pole. It seems that the magnetic field causing these hotspots is a lot more complex than previously assumed[cxliv].

Their super intense magnetism makes the surrounding charged particles swarm around the magnetic poles or "hot spots" and emit the beams of intense radiation, making them look like cosmic lighthouses. These beams will fade away over a few million years as the supply of electrons decreases, but meanwhile their pulses remain accurate to about a billionth of a second per year, about as accurate as an atomic clock. In 1982, the first millisecond pulsars were found, whirling at more than 100 times per second, and one of them (PSR J1748-2446ad) rotates 716 times per second. This is caused by a stellar companion "winding it up" as the neutron star pulls material from its companion. Over time, the neutron star's magnetic field acts as a brake on the star's spin, slowing it down. Nobody was expecting a pulsar that only rotates once every 76 seconds, but that is what we have in the pulsar known as PSR J0901-4046. It's radio-wavelength signal emanates from about 1,000 light-years away[cxlv].

A pulsar that is spinning at 30 times per second powers the Crab Nebula, and that one has a beam that is in the optical range. While some pulsars emit beams in the optical range, and others at X-ray and gamma ray wavelengths, mostly they are at radio wavelengths. A gamma ray photon that came from the Crab Nebula has been calculated as being at 1.1 petaelectronvolts (PeV) or 1.1 million billion electronvolts, making it one of the highest energy gamma rays recorded. For the photon to have that kind of energy it is believed that it must have collided with an electron in the nebula that had an energy close to 2.3 PeV. That would make the electron over 20,000 times as energetic as those in the most powerful electron accelerators on Earth[cxlvi].

PSR B1913+16 is a pulsar discovered in 1974, that is orbiting a non-pulsing neutron star every 7.75 hours. It is pulling constantly on the fabric of the universe, which in turn saps energy from the pair, and the lost energy spreads out as gravitational waves. Observations show that PSR B1913+16's orbit is contracting about 3.2 mm per revolution due to that lost energy, so the two neutron stars should collide in about 300 million years.

I talked earlier about the supernova that occurred in the Large Magellanic Cloud and the energetic X-rays from the remnant of that supernova, SN 1987A, suggest that it is a pulsar, which would make it the youngest pulsar found so far[cxlvii].

Magnetars

Magnetars are slower-spinning neutron stars, but they have the strongest magnetic fields known, and they can release gigantic bursts of energy at irregular intervals in the form of X-rays and gamma rays. All neutron stars may start out as magnetars, although it also appears that the collision of two neutron stars can form a magnetar.

Only 31 magnetars have been identified out of around 3,000 known neutron stars. Swift J1818.0-1607 is estimated to be only 500 years old, revolving once every 1.4 seconds. Also, it is one of the five known magnetars that give off regular radio pulses, along with the X-rays and gamma rays expected from magnetars. These are called radio-loud magnetars, emitting powerful beams of radio waves from their poles. It originally looked pulsar-like but settled down over a 15-day period into a magnetar-like state. Its magnetic poles appear to be substantially misaligned with its spin axis, and they show up just below its equator. Magnetars are seen to be a very diverse group[cxlviii], even when we only have a small sample of them.

On May 22, 2020, a flash from the merger of two neutron stars was seen. The event was called GRB 200522A and it was a kilonova 100 million times intrinsically brighter than the Sun and releasing more energy in a few seconds than the Sun will in its 10-billion-year life. Thankfully, it was 5.5 billion light-years away. The afterglow was ten times brighter than anticipated, meaning that the new object was pumping more energy into the gas cloud, so it seemingly hadn't formed an unstable heavy neutron star that collapsed after a few milliseconds into a black hole as had been expected. A magnetar could inject such energy, but they were not anticipated to form that way, so astronomers are waiting to see if they pick up radio waves within a few years from the ejected material[cxlix]. It is quite possible that this massive magnetar may be unstable and could collapse into a black hole over time. It also seems that a collision between a white dwarf and a neutron star can create a magnetar.

An object 4,000 light-years away and known as GLEAM-X J162759.5-523504.3 is thought to possibly be a magnetar but, if so, it's even stranger than the others. It seems to be spinning remarkably slowly, because it brightens every 18.18 minutes over a span of between 30 to 60 seconds, which is far less frequently and spanning over a much longer period than a normal magnetar. It also shines a lot brighter than expected[cl]. It appears that we still have a lot to learn about magnetars.

FRBs – Fast Radio Bursts

The first fast radio burst (FRB), lasting 5 milliseconds, was recorded on July 24, 2001, but was not recognized as such until 2006. Now we have dozens that have been observed. Modeling indicates that magnetars could be behind them. As a magnetar flares, it is thought that it sends out a blast of energetic particles that gets accelerated by the powerful magnetic field and they ram into material from previous flares, triggering powerful shock waves that produce the FRB effect. In late April 2020, a magnetar, SGR 1935+2154 began blasting out X-rays from 30,000 light-years away near the center of the Milky Way. Gamma rays came later, and finally a fast radio burst. It was the first recorded FRB in the Milky Way and the first one directly associated with a specific object[cli].

FRB 180916.J0158+65 has been tracked to a spiral galaxy similar to the Milky Way nearly half a billion light-years away. The FRB's source is in an outer arm of the galaxy, where stars are rapidly forming. It is only the fifth FRB to be traced to its origin and it is the second repeating one. The other repeating one, FRB 121102, was from a dwarf galaxy 3 billion light-years away, which seems to be a very different environment. Current theory suggests FRBs are neutron stars, and more specifically magnetars, but that may not be the cause of all of them[clii]. It is thought that some of the one-off FRBs (the non-repeating ones) might be from the final collision of a white dwarf and a neutron star that had been a binary pair as they combine and form a magnetar.

Quark Stars, Preon Stars, Boson Stars and Antistars

Neutron stars are not all neutrons, although they have a lot of them, but they have their atomic matter compressed so much that the atomic nuclei are pressed against each other. What would happen if gravity crushed them even more? That might break apart the neutrons and any protons into their constituent components, known as quarks. As such, the star would be called a quark star. It is possible that some of the so-called "down" quarks might turn into their heavier relative, the strange quark, and the mixture of quarks then becomes known as strange matter and the star would be called a strange star. Astronomers have found some stars that they assume are probably neutron stars, but they are smaller than expected, so these may be quark stars and, as we mentioned earlier, there might be quark matter at the core of "normal" neutron stars.

If the mass was even larger, and the gravity stronger, the quarks themselves might get crushed and we would be left with preon stars (preons being the point-like particles believed to make up the quarks that make up protons and neutrons)[cliii]. But no one is really sure if preons actually exist, so preon stars (which might be about 4 miles in diameter and be supremely dense) may not exist.

All of the stars mentioned above are made of the elementary particles that make up normal atoms, namely electrons, protons, and neutrons, or made up of the smaller particles that make up those elementary particles. That is, they ultimately consist of fermions, but there is another type of particle that carries the universe's forces. Those particles are called bosons, and they are ultralight particles that might be just one-billionth the mass of an electron. If there are any low mass, stable bosons out there, they could come together as a boson star. Such stars might have formed in the early days of the universe, and it is speculated that they could be the objects in the center of active galaxies instead of black holes. A 142-solar-mass black hole, detected by its gravitational waves, could have formed through the collision of two boson stars[cliv] according to one model.

Antistars, if they exist, would be stars made of antimatter and would look like normal stars, but would be a source of gamma radiation as atoms of normal matter in space annihilate with the star's antimatter. It is estimated that there could be up to one antistar per 400,000 regular stars in the Milky Way. 14 potential antistars have been identified, based on observations of high-energy gamma rays[clv]. The Alpha Magnetic Spectrometer was bolted to the International Space Station (ISS) in 2011 and it has detected potential signs of eight antihelium atoms during its first ten years, and the simplest way to produce antihelium would be in an antistar.

Black Holes

We talked about neutron stars and other possible types of stars that might result when a giant star collapses, but if the original star had been around 25 or more solar masses (the exact cutoff between neutron stars and black holes is still unclear), the gravity of the remnant that is left after the supernova explosion would be so intense that it would even crush the neutrons, quarks and preons. Then we would have a black hole, and the lightest black hole found so far (XTE J1650-500) contains 3.8 solar masses. The gravity of a black hole will very noticeably bend light beams that pass by it.

When a black hole is created, it distorts space, sending out ripples in spacetime that are known as gravitational waves, and they compress and stretch everything in its path. Their ripples can be so strong that a planet could get ripped apart by them if it was too close, but the strength of the ripples drops off rapidly, which is why we had such a problem detecting them.

Stars with an initial mass of 25 to 40 solar masses are the ones that are most likely to blow themselves up in a supernova and leave behind a black hole, with the remains crushed down to an infinitely small point that is infinitely dense and is known as a singularity. That is assuming that a singularity can really exist, otherwise it would just be something really really small. There is no way you could measure the weight of a teaspoonful of that. A supermassive black hole might be a billion times the Sun's mass, and they are believed to have formed from the collapse of matter in the early universe and have grown mainly by merging with other supermassive black holes when galaxies combined. Stellar-mass black holes are considered to be those that have around five to a hundred times the mass of the

Sun.

So, everything supposedly gets crushed down to an infinitely small point called a singularity, but the singularity itself is not what we normally think of as the black hole, although it is in the center of it. Being infinitely dense means that it has an intense gravitational pull, to the point where light itself cannot escape if it gets too close. That point of no return, even for light, is called the event horizon. The actual "black hole" that we "see" (and have now taken pictures of) is really the event horizon. The distance from the singularity, within which light cannot escape, is also called the Schwarzschild radius, which marks the position of the event horizon. Since no light can escape, the event horizon forms a pitch black "hole" in space, and anything that falls across the event horizon will never come out.

There is speculation that the force from a black hole can even bend spacetime enough to create a wormhole through space and time, and if you could fly through one, you would end up anywhere in the universe, maybe another time (past or future), and maybe even in a different universe. However, calculations show that the wormhole would collapse before you could travel through it, unless you could come up with some form of exotic energy to hold it open. It's more likely that you'd arrive in whatever the "afterlife" is.

There's no actual barrier at the event horizon (at least according to general relativity) and, if you were flying into a black hole, you wouldn't notice it from inside your spacecraft but, once you pass that point, there can be no escape.

You would probably find that you had entered a very violent place as matter was orbiting and crashing into itself as an extension of the accretion disk but inside of the event horizon. That is all assuming that our understanding of black holes is anywhere near correct, and since the laws of physics that we know about, and can confirm, break down in the vicinity of a black hole, we cannot be absolutely certain about anything related to them.

Black holes can have enormous mass, but the event horizon for Sagittarius A* (which holds about 4 million solar masses) is only 15 million miles (24 million km – 17 Sun widths) across and would fit inside Mercury's orbit (although neither Mercury nor the Sun would appreciate it). Stellar-mass black holes with ten times the Sun's mass would have an event horizon about 37 miles (60 km) across. Intermediate sized black holes (with around 10,000 solar masses) would have event horizons about 37,000 miles (60,000 km, or 4.7 Earth widths) across.

The supermassive black hole at the center of the Milky Way is about 25,600 light-years away from us but, even if we were closer, we would not be in much danger of being pulled into it. Mostly, stars and other matter gets pulled into orbits around a black hole, and we see such stars racing around Sagittarius A*. Matter collects into an accretion disk that orbits the black hole rapidly, and there is intense energy and magnetic fields within the disk as the matter collides.

Tidal Disruption Events (TDEs) occur when a black hole tears a star into a stream of hot, glowing gas that is eventually consumed either completely or partially[clvi]. As the star, or other matter, gets sucked into a black hole, it will find itself being stretched and pulled apart. As the star began approaching the black hole, the intense gravity of the black hole would have more effect on the forward part of the star than on the rear (because gravity decreases rapidly with distance), so the star would start to get stretched out, or "spaghettified" to use the technical term, until it was a string of atoms.

Such a star would initially be pulled into an oval shape and then into a long stream of matter. The nuclear fusion in the star stops as the matter density drops and the remains get deposited into the accretion disk, but the stretching and spreading out of the matter causes a brilliant flare reminiscent of a supernova. Some of that matter will get absorbed ultimately by the black hole, although about half of it is likely to end up being ejected as a beam of matter and radiation as gravity and electromagnetic forces spin it up towards the poles of the black hole where the only other way it can go is up and outward. This energetic beam of matter, traveling close to the speed of light, produces radiation in a wide portion of the spectrum including visible light, radio waves and gamma-rays[clvii]. TDEs occur a lot less frequently than supernovae, and it is thought that one may only occur in the Milky Way about every 100,000 years on average.

We see jets of matter streaming out from the black holes at the center of some galaxies, and we have seen stars shooting through our galaxy because they got too close to our supermassive black hole and became accelerated in the black hole's gravitational well and then shot out. On the face of it, black holes appear to be disruptive, and it was thought that black holes in dwarf galaxies would hinder star formation. However, the dwarf galaxy Hen 2-10 has been observed sending jets of ionized gas 500 light-years to the edge of the galaxy, and that expelled gas has become star-breeding regions. So, maybe black holes don't deserve all of their dark reputation[clviii].

Cygnus X-1 was the first black hole to be discovered (discovered in 1964 and confirmed as a black hole in 1990) and it was identified from X-rays traced to a binary system 7,200 light-years away. That system consists of a blue supergiant star and another massive object that had to be the source of the X-rays and it is a stellar-mass black hole. It is now realized that it must contain about 21 solar masses, the largest stellar-mass black hole discovered without the use of gravitational waves, and it makes people wonder if massive stars lose as much material before exploding as previously thought. Since matter falling into a black hole will have angular momentum that is preserved, the black hole will end up with a spin, and Cygnus X-1 is spinning at a good percentage of the speed of light, faster than any other known black hole, and maybe faster than was thought possible[clix].

The singularity at the heart of a black hole is an infinitely small, infinitely dense something. Our universe is said to have come from a Big Bang that began with an infinitely small, infinitely dense something. So, some people believe that a black hole could contain a universe. You could argue that our universe came from an exploding singularity, while a singularity in a black hole is not exploding. But a singularity has a diameter of zero length, so if it expands ten billion times every second, after a year it will have a zero-length diameter, because zero multiplied by any number is zero. Maybe what we see as distance is just an illusion and black holes might be the way that the universe creates baby universes that grow and create their own black hole universes. Hey, here we have another multiverse! It is even seen as a way for universes to become more stable and suitable for life, because it needs a fairly stable universe to produce and sustain black holes. If a universe can produce of lot of black holes, it would increase its number of "descendant universes", so the more stable universes would be the ones that "breed" more and the unstable ones would die out. Here we have evolution's survival-of-the-fittest in action on a multiversal scale.

There are theories related to elementary particles that suggest that there should be ten or eleven dimensions to the cosmos, but we only see three dimensions of space and one of time because the other dimensions are rolled up so small that they are invisible to us. If something moved into a realm with these dimensions, then it would be invisible to us. With black holes, everything is compressed so tightly that maybe all eleven dimensions have the same level of influence there.

After all that is said, black holes (in the traditional sense) may not exist. We know, for instance, that electrons can only take on certain orbits or energy levels. There seems to be minimum energy units, and minimum dimensions that things can have, so maybe it is not actually possible to have a real zero-size singularity. Matter may only crush down to a minimum size, called a Planck length, or perhaps to an even bigger size which might be called a gravistar or a Planck star, but it would still look and behave like the traditional concept of a black hole, so don't lose sleep over it. There is something out there that quacks like a black hole, so we'll call it a black hole.

Primordial Black Holes

We haven't definitely seen primordial black holes, but it is thought that one second after the Big Bang, space may not have been homogenous, with some parts hotter and denser than others. That might have led to some of these regions collapsing into primordial black holes with masses from as low as 10^{-7} ounces (10^{-5} grams) or as big as 100,000 solar masses, and supermassive black holes might have come from the merging of these bigger ones.

LIGO has (at time of writing, late 2021) discovered 47 pairs of colliding black holes, and a statistical analysis suggests that between a quarter and a third (27%) of them might be primordial. That's from comparing the observations with results expected based on models for various kinds of mergers[clx]. However, if Hawking radiation is correct, we might expect to see some of these black holes exploding as they would have nearly evaporated away by now, but we haven't seen that. What we have seen is twenty or so microlensing events that might be explained by primordial black holes. If they are out there, they could explain at least a small portion of the dark matter.

It has been suggested that the surface of the Moon might be a place to look for signs of them. By now, the primordial black holes would have shrunk through evaporation, so you could have one with the mass of an asteroid but the size of an atom smashing through the Moon at maybe 125 miles per second (200 kilometers a second). At its entry and exit points you would expect to see a crater about three feet (one meter) wide with the surrounding ejecta extending a fair distance around the crater and forming a steeper slope than normal. If they are there, it is hoped that they might show up on current high-resolution images from lunar orbiters[clxi].

Supermassive Black Holes

Supermassive black holes can have millions to billions of solar masses and are believed to exist, or to have existed in the past, in the center of all galaxies, except for dwarfs[clxii]. However, M33, the third-largest galaxy in our Local Group, doesn't seem to have a supermassive black hole, but it probably doesn't miss it because supermassive black holes only make up about one half of one percent of the total mass of a galaxy, but they might have been essential to the galaxy's original formation. Also, the Hubble and Chandra space telescopes failed to find a supermassive black hole in the central galaxy of the Abell 2261 cluster, suggesting that a galaxy merger ejected it. There is mounting evidence that black holes can get dislodged from the center of a galaxy as a result of a galaxy merger, and maybe even thrown out of the galaxy.

According to the Lambda-Cold Dark Matter cosmological model, the Big Bang occurred about 13.8 billion years ago, and the early universe was initially a nearly featureless cauldron of subatomic particles and dark matter (with proportions of something like 6:1 for dark to ordinary matter), with everything spread out evenly and unable to clump together into interesting structures. In the first few minutes, lowering of the prevailing temperatures and pressures enabled subatomic particles to form hydrogen and helium nuclei and a faint trace of heavier elements. 380,000 years later, after further cooling, electrons combined with the hydrogen and helium nuclei. That resulted in the formation of neutral atoms and released the cosmic microwave background radiation. The map of that radiation shows that subtle irregularities in the density of matter would have enabled gravity to produce stars and galaxies over time. It was initially thought that black holes grew as massive stars in a galaxy died, and smaller black holes collided and merged until they grew into supermassive ones. But we have found that the earliest galaxies that we can see already have massive black holes at their center, so we are now swinging around to the idea that the black holes could have formed first. Then the black holes might pull in the gas that suffused the universe, wrapping the gas around itself and spinning it out into a disk of gas from which the galaxy's stars formed.

At present we can't see back to when stars and galaxies began forming, so we rely on computer simulations. Those indicate that the clouds of hydrogen and helium gas fragmented into multiple clumps that formed stars with masses equivalent to up to 500 Suns. Those stars shone millions of times brighter than our Sun but only survived a few million years before dying in stupendous supernova explosions and their cores collapsing into black holes with 100 to 200 solar masses. Those black holes then started devouring nearby material and maybe merged with other black holes but could not consume enough in 500 million years to reach a billion times the Sun's mass. That's because matter pulled towards a black hole forms an accretion disk that whizzes around the black hole at near-light-speed, heating up and blasting out radiation across many wavelengths. That radiation pushes away nearby matter, limiting how fast the black hole can add mass. So, we need to find a way to get seed black holes with higher initial mass.

One model shows how dense primordial gas clouds (essentially protogalaxies) in the early universe could collapse to form black holes with masses in the order of 1,000 to 100,000 Suns. A few of the gas clouds might have done that, while most fragmented to form a multitude of massive stars. The difference is in how the cloud cools. Most of the clouds contained an abundance of molecular hydrogen (H_2) consisting of two bound hydrogen atoms which cool quickly, causing them to fragment into clumps that will form stars. However, assume that two or more gas clouds with potential for creating galaxies form near each other in the early universe, and one starts forming stars before the other(s). Then the UV radiation from those stars could affect any adjacent gas cloud, breaking the chemical bonds of molecular hydrogen and converting the cloud into nearly pure atomic hydrogen (H) which is less efficient at radiating away energy, and so it remains hot and can't fragment. Being hot means that the atoms of hydrogen are moving around more rapidly and constantly mixing up the cloud rather than allowing it to fragment. The result is that the gas, that might have become a protogalaxy, collapses as a direct-collapse black hole (DCBH) with maybe as much as a million solar masses. Gravity would then quickly cause it to merge with the nearby galaxy, growing rapidly as it feasted on stars and gas while it moved to the galaxy's center. We have seen what appears to have been a smaller direct-collapse black hole being created. A 25-solar-mass star was seen to "disappear" and it is believed to have collapsed directly into a black hole named N6946-BH1[clxiii].

The supermassive black hole is normally only a few percent, at most, of the host galaxy's total mass, but a DCBH could exceed the combined mass of stars of its small host galaxy by up to fifty times. The supermassive black holes in the local universe typically have only $1/1,000^{th}$ the mass of the galaxy's stars, but small galaxies have been seen to have relatively large black holes. Obese black hole galaxies would be quite bright at infrared wavelengths and have a characteristic spectrum. The James Webb Space Telescope should be able to detect DCBHs at redshifts of 9 to 12 (when the universe was about 526 to 354 million years old), so any quasars with those redshifts would seem to have to be DCBHs.

Another idea regarding forming massive seeds to grow into the early supermassive black holes is for the multiple massive stars to create many smallish black holes. They then plow through the dense gas of the host galaxy and rapidly migrate to the galaxy's center, merging over 50 to 100 million years to form a single black hole of 10,000 to 100,000 solar masses, and then continue growing quickly. These mergers might get picked up on gravitational-wave detectors, such as LIGO/Virgo and the future Laser Interferometer Space Antenna.

It is also possible that some early black holes could accrete matter much faster than is possible now. Starting with a black hole of a few hundred solar masses, it might bulk up relatively quickly to a billion solar masses, and that is thought possible because quasars are eating a lot of material but not emitting nearly as much radiation as expected. Magnetic fields and how they arrange themselves around a black hole affects how matter radiates light as it spirals in. A magnetic field with twists and turns will affect the flow, causing it to heat up and emit powerful radiation, stemming the inflow of material. However, if the magnetic field takes a direct path and neatly threads the accretion disk, simulations show that it can speed the flow of material to the black hole, feeding it much faster than normally expected.

Another model suggests that supermassive black holes grew so quickly after the Big Bang when massive clumps of dark matter at the cores of galaxies collapsed[clxiv], but the universe could easily have used all of the systems mentioned above to achieve its end result.

A supermassive black hole in the elliptical galaxy M87 in the Virgo Cluster is 6.5 billion solar masses (over one thousand times more massive than the Milky Way's supermassive black hole) and an image shows the polarized light from its accretion disk. That indicates that its complex magnetic field could be up to fifty times as strong as Earth's, and it shows that the disk contains highly magnetized gas. The field close to the event horizon is strong enough to push some of the material away, probably accounting for the 5,000-light-year-long jet of material spewing out[clxv]. If you thought that last black hole was big, there is a black hole with 40 billion solar masses at center of the galaxy cluster Abell 85[clxvi].

Midsize or Intermediate Black Holes

There has been ample evidence of stellar-mass and supermassive black holes, but little evidence of midsized ones that range from around 100 to 100,000 solar masses. But on May 20, 2019, LIGO and Virgo detected gravitational waves from the birth of an intermediate-mass black hole. The signal (GW190521) lasted 1/10 of a second and originated halfway across the known universe and was from two large black holes (about 66 and 85 solar masses) merging to create an intermediate-mass black hole about 142 solar masses and releasing around 8 solar masses as energy in the form of gravitational waves[clxvii]. Gravitational wave detectors have been filling in the gaps regarding black holes and showing that they really do come in all sizes[clxviii].

Hawking Radiation

We are now in the Stelliferous Era, the time when stars and galaxies continue to be born. However, the available material will get used up, stars will fade, leaving only black holes. We have seen how black holes grow by devouring the gas and stars around them that get pulled into the accretion disk around the black hole, traveling faster and faster and building up heat through collisions until it glows, outlining the event horizon. Anything that eventually tumbles across the event horizon can't get out, unless you take quantum effects into account.

Although light cannot escape from a black hole because of the intense gravity, electromagnetic radiation (i.e., photons or light) is believed to occur because of quantum effects. The late Prof. Stephen Hawking speculated that the energy of vacuum fluctuations could cause a pair of virtual particles, one with positive energy and one with negative, to be created near the event horizon (as is supposedly happening all over space). Then, one of the particles could get pulled across the event horizon into the black hole, and the other would then escape. Since the one that escapes becomes a real particle, then that one must have positive energy and the one that fell into the black hole must have negative energy, thus reducing the total mass of the black hole by a minuscule amount. So, if a black hole has nothing to feed on, it could slowly evaporate away. The only thing constant in the universe is change.

This Hawking Radiation would be very faint, and nobody has been able to verify that it exists yet, but over time it would cause the black hole to radiate away in the absence of having more matter to swallow. It would take a googol (10^{100}) years for a supermassive star to radiate away. The more massive it is, the longer it will survive. When it gets small enough, the remaining energy escapes in one go, giving a huge flash of light and energy, roughly the equivalent of a million nuclear fusion bombs going off together. That sounds like a big explosion, yet the most powerful supernova recorded (ASSASN-15lh) was 22 trillion times more powerful. Whatever size the black hole started out as, the final explosion will be the same, it just takes longer for that event to come about with massive black holes[clxix].

Scientists have used quantum field theory to show that there should be some form of pressure at the event horizon of a black hole, and that the pressure would be negative, causing the black hole to shrink over time. Whether or not this is linked to the Hawking radiation in some way is not yet determined, but they are both supposedly caused by quantum effects and both make the black hole shrink, so it certainly sounds as if they should be connected[clxx].

It was hoped that the Hawking radiation would solve the "information paradox" or "No-Hiding Theorem" whereby quantum mechanics says that information (that is, quantum information) cannot be destroyed. Quantum mechanics required it to be possible within a closed system to take the current state and extrapolate forward and backward in time to see how the system has changed and will change in the future, yet gravity (through the application of General Relativity) seems to destroy the information and prevent its recovery. It was thought that the particle that escaped from near the event horizon might carry some of that "lost" information back into the universe. However, it has been shown that such radiation is "thermal", i.e., is entirely random and could not be carrying the missing information.

One suggested solution to the information paradox is that black holes are "hairy". By that, it is meant that material swallowed by the black holes leaves an imprint on the black hole's gravity (apart from just increasing gravity due to the additional mass). Apparently, this works with the math for general relativity and quantum theory, but I'll wait and see how the idea pans out when more people have looked at it.

Another interesting point is that the particle that escapes due to Hawking radiation would seem to be connected by quantum entanglement (which Einstein poetically called "spooky action at a distance") to the particle that gets pulled across the event horizon. So, it appeared that there would be some connection between the outside universe and the inside of a black hole, and that idea doesn't sit well with physicists. To get around that, it is speculated that the entanglement might be broken at, or just inside, the event horizon and that the consequent release of energy would lead to an intense "firewall" just inside the event horizon. On the other hand, general relativity says there's nothing special about the event horizon except that gravity there becomes too strong for light to escape[clxxi]. Perhaps we should admit that black holes may be good at swallowing information, but they are really, really, great at generating speculation.

Part 3: Rubedo (Red) – Planets & Life

Chapter 9: Solar Systems and Planets

Alchemical Step 3: Rubedo or Red

So, by this time in the universe's evolution we have all the atoms and molecules that the universe needs, but where does life fit in to this? The alchemists were interested in extending life, but the universe had to actually create it. That takes us to the next step in the astronomical alchemical process.

The ultimate goal of alchemy was to create a means for

achieving eternal life, via the Philosopher's Stone. That final step was called rubedo, or red, implying the full glow of wisdom and initiation, and it is no coincidence that this is the color of blood, the life force. The universe might have its own goals related to eternal life, but we can see it producing life that has its own ways of continuing and developing as long as the universe lets it.

Our Solar System

We have reached the stage in the development of the universe where the stars and galaxies exist or are forming, and there are enough elements to allow more solid bodies, such as planets, to form, which give good places for what we know as life to start developing.

Our Sun, which we should properly call Sol, formed in a giant molecular cloud, which probably had gas equivalent to at least ten thousand solar masses. That cloud might have been around for billions of years before it started to collapse and form an open star cluster. A supernova's shockwave from light-years away could have triggered the collapse of the giant cloud or nebula. Another idea is that the Sagittarius Dwarf Elliptical Galaxy, which is about ten thousand light-years in diameter and is now about seventy thousand light-years away from us, might have triggered the collapse. There seems to be coincidences between its closest approaches to our Milky Way galaxy and major periods of star formation, and an approach 5.7 billion years ago may have resulted in the birth of our Solar System. A third idea is that the stellar wind from a massive Wolf-

Rayet type of star might have started the collapse. That is considered to be likely because the proportion of Aluminum-26 (Al-26) to Al-27 found in meteorites is about 17 times higher than that seen in the Milky Way generally, so something injected the extra Al-26 into the gas cloud that formed us. Meteorites also show far less Iron-60 than in the Milky Way as a whole, and Wolf-Rayet stars could account for both of these anomalies. Wolf-Rayet stars are O-type stars with more than 25 times the mass of the Sun that are near the end of their lives and have ceased normal hydrogen fusion. Being extremely hot, maybe 54,000°F (30,000°C), their stellar winds can exceed 4.5 million mph (7.2 million km/hr.) which could blow and compress the gas cloud while altering its molecular mix. Maybe a combination of these possible triggers was involved, because the universe doesn't like playing favorites, and one study (looking at the effects of supernovae on star birth) indicated that two such explosions worked a lot better than just one[clxxii].

Within that collapsing cloud, a patch that was denser than the large cloud as a whole and was maybe 600 light-years in diameter, containing the mass of around ten Suns, collapsed sufficiently for our Sun to be born at its center about 4.6 billion years ago (September 1 on our "cosmic year" scale). The hydrogen and helium atoms that made up most of the cloud began to pull closer together and, after about ten million years, it had gathered at the center of the cloud where gravity has its focus. As the mass concentrated by the gravity increased at that point, the pressure and heat rose[clxxiii]. It was now a protostar and would stay that way for about 500 thousand years as it continued to pull more matter in and, as the gas fell inward towards it, the angular momentum resulted in the gas cloud becoming a disk revolving around the protostar. When the core of the protostar reached 9 million degrees F (5 million degrees C), nuclear fusion started and it became a real star, and a balance quickly developed between the inward pull of gravity and the outward push of radiation from the fusion, at which point the core was around 27 million degrees F (15 million degrees C).

The Sun was most likely born in an open cluster that dispersed over time, and it may have started out closer to the center of the Milky Way and migrated outward as the spiral arms formed. Interaction between stars in a cluster are relatively common, and the Sun presumably got flung out of the cluster in a gravitational interaction with one of the other young stars. The stars known as HD 162826 and HD 186302 are believed to be two of our solar siblings, the former being found in Hercules and is about 110 light-years away, and the latter is in the constellation Pavo in the southern sky and is about 184 light-years distant from us. Nowadays, the Sun sits in a relatively calm area of the Milky Way, which seems to be much more conducive to life.

But the Sun wasn't left totally on its own. It might have been thrown out of its home, but it had its babies that were forming in a 9 billion mile (14 billion kilometer) wide disk of gas and dust surrounding the Sun. Electromagnetic forces probably pulled the tiny grains of matter together initially, but by the time that a blob of matter had grown to almost half a mile wide, there would be sufficient gravity to pull more matter in[clxxiv]. A few million years after the Sun formed, there were solid bodies growing out of the cloud of gas and dust around the Sun, growing to be asteroid- and comet-sized protoplanets. Some of those protoplanets smashed together and grew into larger bodies like our Earth and brought with them the chemicals that had been blasted out into the universe by supernovae and colliding neutron stars.

One problem that people have seen with that "steady growth" idea is that as smaller bodies (accumulations of dust) get big enough, they might start bouncing off of one another instead of merging and growing. To get around that, it has been suggested that alien bodies, such as 'Oumuamua, might have been captured into an orbit around the Sun and become the core body onto which other matter could accumulate[clxxv]. That might indeed have happened, but it just pushes the problem further back, because we then have to explain how that alien body grew to the size it was. We have already seen objects, such as comets and Oort Cloud bodies that are obviously accumulations of two or more other bodies that came together relatively gently, and we have also seen planets forming in the disks of dust and gas around other stars, so there seems to be no desperate need for alien bodies to start the process.

But that process of planetary creation can certainly be a very dynamic one[clxxvi]. Within the disk of dust and gas, clumps of matter start to grow into planets, but they are affected by the remaining dust and the other growing bodies, so an individual body can end up falling inward towards its star or get thrown outward and maybe out of the system completely. All that movement along the same orbital plane can also result in collisions, shattering protoplanets or causing them to merge. We see the end result of such movements, for instance, in the hot-Jupiters that we find in other solar systems, and we'll see that our Jupiter went wandering and probably caused many of the major collisions we find indications of, including the collision that is believed to have given us our Moon. With all that dynamic action going on, a small change in conditions could have resulted in a much different solar system. Planetary creation is a chaotic system that is nearly as unpredictable as another chaotic system – our weather. It also seems that stars similar to our Sun do not necessarily grow steadily. Matter accumulates within the star in bursts, which leads to rapid brightening at times and substantial changes in the temperature that the star subjects the surrounding matter to, which can alter the structure of the objects forming there[clxxvii]. It's also likely that the Sun's protoplanetary disk and the planets forming in it were affected by other star systems in the stellar nursery where the Sun was born.

Our Solar System consists of the Sun and all of the objects that orbit it. The largest and most obvious of these objects (after the Sun) are the planets, and they can be divided into the terrestrial planets and the gas giants. A common unit of measure that we use when talking about the solar system is the AU or Astronomical Unit, which is the average distance from the Sun to Earth (i.e., it is the average radius of Earth's orbit) and is defined as 92,955,807 miles (149,597,871 km). The terrestrial planets include Mercury at 0.4 AU from the Sun, Venus orbiting at 0.7 AU, Earth is at 1 AU of course, and Mars is 50% further out at

1.5 AU. Beyond the asteroid belt there are the gas giants, including Jupiter at 5.2 AU, Saturn at 9.5 AU, Uranus at 19 AU, and Neptune is way out there at 30 AU.

There are also dwarf planets, which are spherical bodies but with a mass too low to significantly perturb planetary orbits. Generally speaking, a body has to be getting on for 500 miles across to generate gravity sufficient to pull itself into a sphere. We also have small solar system bodies that are not spherical, such as most of the minor planets (more commonly known as asteroids), comets and most Kuiper Belt objects or TNOs (Trans-Neptunian objects). Some of the TNOs are also dwarf planets, including Pluto, Eris, Haumea, and Makemake. Most, though not all, of the asteroids or minor planets orbit in a "belt" between the orbits of Mars and Jupiter. Sometimes, the term "asteroid" is used to include the bodies orbiting between Mars and Jupiter (and which sometimes even travel in our direction), and also the bodies in the Kuiper Belt, but I'll generally restrict the term to objects originating in the "asteroid belt". What is normally known as the Asteroid Belt can be found between 2 and 5 AU from the Sun. There is one dwarf planet, namely Ceres, within this belt. Any of these objects can also have their own satellites or moons, which are smaller objects that orbit them. The Kuiper Belt spans from around 30 to 50 AU and is where most of the short-term comets come from, as well as where Pluto, and a number of other dwarf planets and other smaller bodies, live.

The heliosphere ends at around 123 AU and Voyagers 1 and 2 have now passed that point and Voyager 1 has reached to about 152 AU. The Oort Cloud is where long-term comets come from and it ranges from around 1,000 to 100,000 AU. For comparison, our next nearest star, Proxima Centauri, is 268,000 AU from the Sun. The atoms in the desk in front of me are close up against each other and if you looked at the outer shells of the atoms, you would find they consist of one or more electrons. The outer layer of our solar system reaches almost halfway to the next nearest star, and just as an electron from one atom can sometimes be shared with an adjacent atom, it is believed that bodies in the Oort Cloud of our solar system might sometimes be captured by an adjacent solar system, and we might have stolen similar comet-like bodies from them.

The Sun

The Sun sits at the center of the Solar System and is our closest star, and it is big (compared to Earth, anyway), almost 870,000 miles (1,400,000 km) in diameter. It would take about 109 Earths lined up side-by-side to span the Sun's width. It is also ten times wider than Jupiter, and you could fit 900 Jupiters in it[clxxviii]. The Sun contains almost 99.9% of all the mass in the Solar System and it is very hot. The Sun's surface temperature is about 5,800 kelvin (about 5,500°C or 10,000°F), while the temperature at the Sun's core is 15.7 million kelvin (about 28 million deg. F) and, although it is on the small side as far as stars go, it still produces more energy in one second than mankind has ever used.

At around 8 minutes 19 seconds away from us when traveling at the speed of light, the Sun is moving at close to 135 miles per second (220 km/second) around the center of the Milky Way. The Sun revolves once on its axis every 25.5 days, approximately, at its equator, but slower at the poles, and it has a strong and constantly changing magnetic field which is the cause of sunspots and most of the other solar activity.

Basically, the Sun is a ball of plasma with about 75% of it being hydrogen, and the remainder is helium with a small number of heavier elements mixed in. It has something like a thermonuclear reactor at its core, fusing hydrogen into helium. I have seen car-stickers with a smiley-faced Sun and wording about "No nuclear power", and I can see why the Sun is laughing on those car stickers, since it is the biggest nuclear power source we have around here. The Sun's fusion reaction has been active for about 4.6 billion years and will continue for another 3 billion years or more. While its core is around thirteen times denser than lead, it consists of compressed gas or plasma with the atoms getting squeezed together and fusing. Over time it will have used up most of the hydrogen in its core and will then start fusing helium which will heat up the Sun even more and make it puff up into a red giant.

Outside of the Sun's core is the Radiative Zone where energy travels outward as radiation or photons, although that radiation is constantly interrupted by interaction with the dense matter. Then we get to the Tachocline, which is the border between the radiative and convection zones. The Sun spins rather like a solid below the tachocline and like a fluid above it, and the intense shear at that point helps to create the magnetic field. Because of the density of the Sun, a photon of light can take an average of 100,000 years to reach the surface, and maybe a million years, as it effectively bounces back and forth off the dense matter in the Sun's core. Happily, that wears down the energy of the photon so it's no longer in the range of gamma-rays that can tear atoms apart, and instead the photons are reduced to the frequencies that include visible light and heat (infra-red). Above the Tachocline is the Convection Zone where energy is moved by convection, and the Meridian Flow is a current of plasma within the convection zone that transfers the heat to the surface of the Sun. What we see (through very dark filters, hopefully) as the Sun's visible surface is named the Photosphere. Some odd waves or vortices have been observed moving in the opposite direction to the Sun's rotation and traveling about three times faster than other similar waves. What can be causing these vortical waves remains a mystery[clxxix].

The Chromosphere is the layer of the Sun's atmosphere above the photosphere (the visible "surface"), and the Corona is the Sun's outer atmosphere and, surprisingly, it is about twenty times hotter than the visible surface of the Sun. However, the corona is also very diffuse, so if you were somehow floating in it, and also protected from the radiation streaming out from the Sun's surface, you would freeze to death, not fry, because the intensely hot atoms are spaced too far apart to transfer heat fast enough to compensate for your body radiating heat away. The corona is what makes a total eclipse such a memorable sight, and it expands out into space and creates the solar wind that extends about 100 times as far out into space as the Earth is from the Sun. The solar wind reaches far out beyond the outer parts of Pluto's orbit, to about 11 billion miles (around 18 billion km) from the Sun, and the two Voyager spacecraft have now gone past its furthest reaches, known as the heliosphere. Here the solar wind meets the interstellar medium and it has been found that this region is extremely active as those outer layers of the solar wind protect the solar system against incoming interstellar and intergalactic cosmic rays.

The Sun's magnetic field is about twice as powerful as Earth's, with global magnetic field lines exiting at the positive pole and reentering at the negative pole. These magnetic field lines get pulled into an east-west rotation and become twisted as the Sun rotates faster at the equator than at the poles. The Parker Solar Probe identified features called switchbacks that are dynamic S-shapes in the magnetic field that funnel the plasma and dissipate their energy before disappearing near the top of the corona. It is thought that that might be the source of the heat that makes the corona far hotter than the photosphere, and also what might accelerate the solar wind, rather like waving a whip[clxxx], then cracking the whip to release the energy. Solar Prominences are created by magnetic fields that suspend an arch of gas far above the Sun's surface and if two such magnetic field lines cross, they can short-circuit and create a hundred-million-degree Fahrenheit (55.5 million degrees C) blast. Sunspots are the cool regions on the Sun's surface, but they are still hotter than 7,500°F (4,150°C), and can be around 10,000 miles (16,100 km) across, bigger than the Earth, and sometimes up to twenty times as big. They are dark spots where amplified magnetic fields, that are anchored far within the Sun, break through the surface. Carried by the deep meridional flow, sunspots emerge closer to the equator as the solar cycle progresses. Solar flares are sudden releases of energy stored in sunspot magnetic lines[clxxxi]. Also, something like miniature solar flares, called "campfires", have been seen, and they are similar to the nanoflares that could be another source of heating for the Sun's outer atmosphere, or corona[clxxxii]. The Solar Wind is a thin ionized gas that is sent speeding away from the Sun. Coronal Mass Ejections are billion-ton clouds of charged particles that leave the Sun moving at around 1,800 miles per second (2,900 km/sec) and can last for days. The Sun's magnetic field produces some very dynamic events, and the field remains surprisingly complex far from the star.

As the Sun ages, about 3 billion years from now it will start growing in size as it begins fusing helium, making it much hotter. It will then burn off so much hydrogen that there will only be a shell of hydrogen left around the helium core that is fusing helium into oxygen and carbon, turning the Sun into a red giant, initially disrupting the orbit of Mercury and maybe sending our smallest planet back out in our direction. Then the outer layers of the Sun will engulf Venus, making it even hotter than it is today. The Sun could even grow as big as Earth's orbit. But humans (in the form of Homo sapiens) have only been around for about 200,000 years so far, and it will be about 5 billion years before Earth is in danger of being swallowed by the Sun, and by then we might be able to move the Earth, or at least its inhabitants, to a safer location. The Sun is currently increasing its radius about an inch (25 mm) a year on average while the Earth is expanding its orbit around 6 inches (15 cm) per year, so it looks as if our planet has started taking its own action.

We have discovered one planet on which any oceans will have been boiled away as its star expands and grows, and that planet will soon be swallowed by the red giant star. But it seems that that doesn't necessarily mean the end, because we have also discovered a planet that has reemerged from a shrinking red giant. Okay, any life on the planet would be long gone, but some remains of the planet survived.

As the Sun continues heating up, it will be blowing off its outer layers until finally the Sun won't have the gravity to maintain the temperatures for future fusion and it will collapse down into a white dwarf.

In astrology, the Sun is the ruling "planet" of Leo, and it is also exalted in Aires. It travels along the ecliptic through each of the zodiac's signs during the course of a year, and its position at your birthday defines your birth sign, although that relates to the astrological "house" rather than what we know as the zodiacal constellations. The Sun supposedly represents your ego, personal power, and pride and authority, which all adds up to represent your "life force", and the Sun certainly provides basically all the power we use here on Earth (with the notable exception of nuclear power, but other stars provided the nuclear fuel that we use). It supposedly governs bile temperature, the stomach, bones and eyes (it can certainly affect your eyes if you don't take proper precautions when looking at it).

Ruling Planets

When astrologers say that a planet "rules" your birth sign, they imply that the planet's energies are supposed to be the primary influencer for that sign. For instance, Mars supposedly makes you a risk-taker, assertive and aggressive, which helps make those under Aries be motivated and confident leaders, and those under Scorpius to be tough-minded and biting. However, a birth sign has other supposed influences too, such as the traditional elements of Fire, Water, Earth and Wind, so each of the astrological "zodiac" signs end up with its own unique set of aspects.

Planets are also described as being "exalted" in certain signs, and that means that any influence that the particular planet has for someone with that sign will become enhanced.

What Is and Is Not a Planet?

The reason the Sun was called a planet by the ancients is because the word "planet" originally came from the Greek word planete which means "wanderer". Like the Moon and what we now normally call planets, the Sun appears to wander across the sky, unlike the other stars that appear fixed.

Some of the other moons of our solar system have been

called planets in the past as well. When Galileo first discovered the large moons of Jupiter, they were called planets, largely for want of a better name. Saturn's moon, Titan, has an atmosphere, clouds, rivers and lakes, making it very much like our planet, although those lakes and rivers are liquid methane, not water.

The modern definition of "planet" says that it must be nearly spherical, and it must orbit its star (Pluto meets both of those requirements) and it must "dominate" or basically control anything in its orbit.

That definition of a "planet" runs into some difficulties with regard to "dominating" an orbit, particularly with exoplanets where we have enough problem ascertaining if the planet is there, let alone what else is happening within its orbit. And how do you deal with rogue planets that no longer orbit a star or anything at all? One suggestion has been to replace that final portion of the definition with a requirement that the body is, or has been, geologically active[clxxxiii], which would let Pluto back into the fold if it was adopted. Anyway, it's currently a reasonable, if not perfect, definition for "planet".

Planets

When our Solar System was being formed, the gas cloud started collapsing and the Sun burst into light as the pressure at its center started the fusion of hydrogen atoms. The rest of the gas cloud settled into a disk revolving around the Sun's equator, and the planets grew initially by dust in the gas cloud around the Sun collecting into "dust bunnies", then into larger accumulations that become small rocks, then those rocks came together into larger rocks, growing into protoplanets that were rather like the asteroids of today. Where the gas giants are concerned, that "dust" might have been mostly ice crystals. An alternative idea is that portions of the protoplanetary disk became gravitationally unstable and collapsed directly into protoplanets, in much the same way that stars are believed to have formed[clxxxiv]. However they arose, there were a lot more of these protoplanets than there are asteroids nowadays, and they collided, combined, and grew into the planets that we now know. We even see evidence of some of those collisions today.

We think of planets moving in circular orbits around their star, but in fact they follow elliptical orbits with the star (the Sun in our case) being at one of the focal points. For objects orbiting the Sun, perihelion is the point on the orbit closest to the Sun and aphelion is the point on the orbit furthest from it. The closer to the Sun that a planet orbits, the faster it moves, and the speed varies in inverse proportion to the distance from the star (the further away they are, the slower they will be moving). Planets and their orbits are also affected by the gravity of other planets and, when the solar system was young, by the gas cloud that they formed from. Since that had not been used up, the gas could slow down planets and change their orbits, transforming the solar system.

The bulk of the gas cloud disk formed one large planet, that we now know as Jupiter. But there was still a lot of the gas left, which slowed Jupiter down in its orbit and it started to spiral in closer to the Sun. As it moved, it disrupted the other planets that were forming closer in, and they were already colliding with one another. Our Moon is believed to have formed when a Mars-sized planet (that we call Theia) collided with Earth and some of the matter from the collision got thrown into Earth-orbit and pulled itself together into the Moon. Mercury seems to have started out in an orbit closer to us, but it collided with something, maybe Venus, losing most of its crust and leaving not much more than its metallic core. The resulting small planet dropped into its orbit not far from the Sun. Such a collision could also account for Venus's odd spin. Impacts on Mars threw surface material out into space, and we have found some of it that has fallen on Earth in the form of meteorites. It appears that something very big even knocked Uranus over, and it is now almost lying on its side as it orbits the Sun.

But Saturn formed soon after Jupiter, and its pull stopped Jupiter's inward spiral, and began dragging our giant planet back out again. All these movements of the gas giants caused havoc in the Solar System. Uranus and Neptune appear to have ended up swapping positions, and one possible gas giant might have gotten itself almost thrown out of the solar system and could be what we currently call Planet 9.

Evidence for Jupiter's ramblings, and consequent disruptive/transforming effects, include the small size of Mars (presumably because Jupiter disrupted its potential source material), and the distance of the ice giants from the

Sun (which seems to be further out than where massive planets could have formed from what would have been the thinning outer edge of the protoplanetary disk). Also, two groups of small bodies formed in the solar system, the first being the asteroid belt inside the orbits of the gas giants, and they are made up of non-carbonaceous material (i.e., they are low in carbon). The second is the Kuiper Belt which is outside of the gas giants' orbits, and the objects there are carbonaceous (with higher levels of water and carbon that survive better further out from the Sun). Indications that these separated belts got mixed up by some event has been found in a meteorite that fell in India in 1870. Known as the Nedagolla meteorite, it contains elements corresponding to carbonaceous and to non-carbonaceous meteorites. Radioactive dating suggests that the original meteoroid formed at least 7 million years after the birth of the solar system as bodies from the two regions collided, probably as a result of Jupiter's movements[clxxxv].

Since the planets all formed from that same revolving disk of matter, they all lie approximately on the same plane. Earth is perfect, of course, and lies exactly on the plane of the solar system, but maybe that's because we define that plane as being an extension of the line running through the centers of the Sun and Earth. The deviations of the other planets from this plane of the solar system are as follows[clxxxvi]: Mercury 7.0°, Venus 3.4°, Mars 1.9°, Jupiter 1.3°, Saturn 2.5°, Uranus 0.8°, Neptune 1.8°, and Pluto (let's not forget everyone's favorite dwarf planet) is 17.2°.

Inner or Terrestrial Planets

The terrestrial or inner planets are the rocky planets that orbit inside the asteroid belt and include our Earth, although we won't be discussing Earth specifically in this chapter. In order from the Sun, these inner planets are named Mercury, Venus, Earth, and Mars. Interestingly, the mass of the Earth is slightly larger than the mass of the other three terrestrial planets (Mercury, Venus and Mars) combined. All four inner planets were made from the same ingredients, but they followed very different evolutionary paths, and that made each of them very interesting in its own right. For instance, if we define the standard for atmospheric pressure as being Earth's air pressure at sea level[clxxxvii] and call it 1, then Mercury's atmospheric pressure is 0, or as near to it as makes no difference, Venus's is 92, and Mars has an atmospheric pressure of 0.01.

Sometimes the term "inferior planets" is used to refer to Mercury and Venus. "Inferior" does not mean "substandard" or in any way "unworthy" but simply refers to the fact that their orbits are inside of Earth's orbit. Under that type of terminology, Mars is included with the gas giants and together they are referred to as "superior planets", meaning their orbits lie further away from the Sun than Earth's orbit. Until getting demoted to a dwarf planet, Pluto was a "superior planet".

Mercury

Mercury is named after the Roman god of the same

name, who was the messenger of the gods and was known for his speed and swiftness. In astrology, it is the ruling planet of Gemini and Virgo, and it is also exalted in Virgo. The planet stands for communication, thinking patterns, reasoning, adaptability and variability, in other words being quick witted (being the closest to the Sun, it is the planet that moves the quickest in its orbit). It governs schooling and education, and it rules over Wednesday.

Mercury has a cratered surface, looking much like the Moon but it has no moons of its own. It is the first planet from the Sun, about 3,030 miles (4,880 km) in diameter, just under 40% of Earth's width and it is not a lot bigger than our Moon. It is the smallest of the planets now that Pluto has been demoted and it has a mass that is only about 5.5% that of Earth.

It orbits the Sun at a distance of around 36 million miles (58 million km), about 40% as far away from the Sun as Earth is, although its orbit is the most eccentric of all the official planets and, as mentioned above, it travels at the fastest speed of all the planets. Its orbit takes about 88 of our days, but its day (sunrise to sunrise) is twice as long, so it takes about one of its years to get through its daylight hours and one "year" to get through the night. Its real rotation period against the background stars is 58.8 days.

Its surface gravity is about 40% of Earth's or double that of the Moon and, like Earth, it has a magnetic field. Being so close to the Sun, it gets very hot, and Mercury's surface temperature ranges from around minus 280 degrees F (-175°C) on the dark side to plus 800 degrees F

(430°C) on the sunny side, giving Mercury the widest temperature swing of any of our planets.

Being a terrestrial planet, it is rocky like Earth, and it is believed to have a higher percentage iron content than any other planet in our solar system (maybe 75% of the planet's mass), but it has no atmosphere worth talking about, although the little that it has is 42% oxygen, 22% sodium, 22% hydrogen, 6% helium, 8% other gasses. The solar wind causes the sodium and other particles to blow out from Mercury in a faint glowing comet-like tail stretching around 25,000 miles (40,000 kilometers) from the planet[clxxxviii]. Surprisingly, it is believed that ice may exist in some sheltered craters near the poles, where sunlight never reaches. It was thought that the ice might have come from comets or asteroids, but it is also possible that at least some of it might be a direct result of the Sun. The idea is that the protons blasted out from the Sun in the solar wind break down some of Mercury's rock releasing oxygen that combines with hydrogen to become water in the form of ice[clxxxix].

The largest crater on Mercury has been named Caloris Basin and is about 950 miles (1,550 km) in diameter, which is over half the diameter of the planet itself. That crater is believed to have formed around 3.8 billion years ago when a 93-mile (150 km) diameter body crashed into the planet. That sent seismic tremors through the planet disrupting the surface on the opposite side, resulting in an area that astronomers describe with the technical term "weird"[cxc].

Compared to our Moon, there are very few boulders on the surface of Mercury. It is thought that this might be the result of the intense heat from the Sun degrading any boulders, or the surface of Mercury having a thicker layer of dust on its surface than the Moon has, resulting in fewer boulders being broken off by meteor impacts[cxci].

Venus

The symbol for Venus is the same as that for females and that symbol is also sometimes said to represent the mirror of the goddess. Venus is named after the Roman goddess of love and beauty, and probably because it is so bright it is known for charm and allure. It is the ruling planet of Libra and Taurus, and it is exalted in Pisces. The planet is associated with harmony, beauty, sympathy, and affection (not reflected in its current climate). It governs relations, from business to romantic, and it rules over Friday.

Venus is our closest neighbor, and it is considered to be a sister planet to Earth and, billions of years ago, it may have been a lot more like Earth[cxcii]. Its core is about 4,400 miles (7,000 km) across, very similar to Earth's, and it may have had water on the surface with rivers and oceans and the ingredients to support life-as-we-know-it. But hopes for the planet harboring such life now on its surface were destroyed by the first successful unmanned spacecraft to visit the planet, Mariner 2, which showed how hot the surface was. Before that, our space agencies had been planning for possible splashdowns for any unmanned spacecraft landers on Venus. Now we know that Venus has a scorching hot and dry desert-like surface with volcanoes.

Its highest mountain is called Maxwell Montes, 7 miles (11 km) high, which is one of the few features not named after a woman. Some of its mountains may look as if they have snow on them, but it's metallic snow. Metals, such as bismuth and lead, melt, evaporate in the heat and fall like snow[cxciii].

Venus is the second planet from the Sun, about 7,500 miles (12,100 km) in diameter which is almost 250% as big in diameter as Mercury, and 95% as big as Earth. It orbits at around 67 million miles (108 million km) from the Sun, which is about 70% as far away from the Sun as Earth is. Its orbit is almost perfectly circular and the planet "laps" us every 584 days. Its actual orbit (i.e., its year) takes about 225 of our days. It rotates once every 243 Earth-days, plus or minus 21 minutes, in relation to the background stars, but in the opposite direction to every other planet, except Uranus, with the result that its apparent solar day is only just under 117 of our days. The variation in length of its rotation is due to the thick atmosphere that pushes and pulls on the surface. Its atmosphere near the surface is so thick and dense that it's more like a liquid, but that means that wind speeds are down to about walking pace (4 mph or 6.5 km/hr) while, about 35 miles (55 km) up, in the lower reaches of the clouds, the wind clocks in at about 300 mph (480 km/hr)[cxciv]. The planet wobbles, as a result of the Sun's pull, in a way that might replicate about every 29 thousand years. Like Mercury, Venus has no moon.

Venus is the hottest planet in our solar system, averaging about 860 degrees F (460°C) at the surface and that would melt lead, so it is even hotter than Mercury. Its runaway greenhouse effect is thought to have been started by the Sun getting hotter and beginning to vaporize the oceans. That caused the atmosphere to become filled with water vapor which is a potent greenhouse gas. The result was the retention of more heat, and it finally left the planet with its reflective clouds of sulfuric acid. Lightning has been seen in the clouds, and the clouds produce sulfuric acid rain that evaporates in the searing heat long before it hits the planet's surface. Venus may be a lady but she's a fiery one.

One model of Venus's formation suggests that she always was a fiery one and was always too hot for liquid water to have been on its surface. That idea has Venus being born as a steam world, with the only water being in the form of water vapor in the atmosphere. There may still have been regular clouds that allowed water rain to fall, but the raindrops would have vaporized in the heat before hitting the planet's surface, just as the sulfuric acid rain does today[cxcv]. On the other hand, the first probe to land there in 1967 saw rocks that looked like granite, and that requires water for the rock to form.

The most obvious feature of Venus, and the only part that we can see from Earth, is its sulfurous clouds that cover the whole planet and keep the heat in. But the clouds are not that hot themselves. The cloud cover spans from about 30 miles (48 km) above the planet's surface, up to about 43.5 miles (70 km). At the bottom of the clouds the temperature is around 230°F (110°C) and the pressure is about twice the pressure we experience on Earth at sea level. At the top of the clouds, the temperature is down to around 113°F (45°C) and the pressure is only 4% of the pressure at sea level here, so the clouds would be considered a habitable region for any Earth microbe that likes the sulfuric acid droplets of the clouds. The cloud layer seems to rotate once over a period of about four days, so it spins about 60 times faster than the planet itself[cxcvi].

Hundreds of active volcanoes in the past also added greenhouse gasses to the atmosphere, although we don't think that any of the volcanoes remain active. Venus does not have plate tectonics like Earth does, but it appears that there are circular features called coronae or campi which might be as big as Alaska and are believed to form when faults develop around areas where rising magma lifts up the surface, and it seems that at least 37 of the planet's large coronae are still evolving, showing geologic activity. These blocks appear to get shifted around as convection within the planet's viscous interior moves the crust.

Venus's atmosphere is now almost exclusively carbon dioxide, plus 4% nitrogen, and less than 0.1% other gasses. Its atmospheric pressure is over 90 times that of Earth at the surface. Because of its highly reflective clouds, it is the brightest natural object in our sky after the Sun and Moon. It is another terrestrial, rocky planet, but unlike Earth and Mercury and despite its solid inner and molten outer core, it does not have a noticeable magnetic field of its own, only a weak induced field resulting from interaction with the solar wind. It presumably lost its own magnetic field when the planet was hit by a planetary-sized object which slowed its spin, effectively stopping the generator in its molten core from operating.

Mars

Mars is named after the Roman god of war, it represents masculinity and youth, and its symbol is the same as the one that represents men. Mars is the ruling planet of Aries and Scorpio, and it is exalted in Capricorn. The planet stands for confidence, self-assertion, energy, even aggression (even though it is small and basically dead). It governs sports, competitions, and physical activity generally. It rules over Tuesday, which for French, Spanish or Italian speakers is more obvious than it is for those of us who only speak English (e.g., Tuesday in Spanish is Martes).

Mars, the Red Planet, is the most distant of the inner planets from the Sun and our second-closest neighbor, and it has a cold desert-like surface with volcanoes, valleys, and signs of water flows in the past. Mars's red coloring comes from the iron oxide, or rust, on its surface. It is thought that most of the volcanic eruptions on Mars occurred between 3 billion and 4 billion years ago, with some occurring up to around 3 million years ago. There is even some indication that volcanism may have occurred at Cerberus Fossae as recently as 50,000 years ago. It has quite a few craters, including one truly massive one.

Borealis Basin is a crater as big as Africa and covering about 40% of the Martian surface and is believed to be the result of an impact by an asteroid estimated to have been about 1,200 miles (1,930 km) wide[cxcvii], almost as big as Pluto. That would have sent vast portions of the crust into space and sterilized the planet, just as Theia did to Earth, but it probably only took Mars about fifty million years to recover[cxcviii] – the blink of an eye in cosmic time. Mars's two small moons may have been formed by such an impact, although it's still not 100% definite that the Borealis Basin is an impact crater.

For a small planet, it has some very large features including the Borealis Basin, of course, and also vast canyons that can run for distances equivalent to that from New York to Los Angeles, and despite having little in the way of an atmosphere it can have sandstorms covering nearly the whole planet that can last for months.

At one time it was believed that Mars had a thriving civilization that had built canals, but the canals are now known to have been an optical illusion and, if it has any native lifeforms, they are not much more than microbes. At one time it did have rain, rivers and seas, and organic compounds have been discovered in the soil, giving hints that life might have thrived there once. Its seas were probably covered by ice, but with the low gravity there could also have been waves far higher than those on Earth.

Mars may not have canals, but it is crisscrossed with old riverbeds, a small proportion of which seem to have been gouged out by catastrophic flooding. These deep ravines make up only about 3% of the observed ancient river flows, but they are much deeper than the others, averaging about 560 feet (170 meters) deep. It is estimated that about 25% of the Martian soil and rock cut away by water flows came from these types of canyons[cxcix]. Rocks and boulders up to 5 feet (1.5 meters) wide have been observed that were moved by a river that flowed into Jerzero crater, which shows the kind of intense flow that must have happened. The topography of the outlet canyons appears to indicate that there was some massive flooding about 3.5 billion years ago[cc] (September 30 on our "cosmic calendar").

Mars's largest mountain, Olympus Mons, makes Everest look like a little hill, being almost three times as tall at 17 miles (27 km) high by 300 miles (480 km) across its base, but it's not actually the tallest mountain in our solar system, although it is the tallest on any of our planets. The central peak of the Rheasilvia crater on the asteroid Vesta stands slightly higher, base to peak, than Olympus Mons.

Mars is the fourth planet from the Sun (we skipped over the third planet, Earth), and it is about 4,220 miles (6,790 km) in diameter, which is 53%, or a bit over half of Earth's radius. That makes it the second smallest of the planets with only about 11% of the mass of Earth. It orbits at around 142 million miles (228 million km) from the Sun, ranging from 129 million miles (207 million km) to almost 155 million miles (250 million km), around half as far again away from the Sun as Earth is. The terrestrial planets had been getting larger as we moved out from the Sun, so why is Mars so small? As we mentioned earlier, Jupiter started rambling through the solar system early in its life, moving inward to around where Mars is now, and that probably resulted in there being little material left to form much of a planet there.

Mars's orbit takes about 687 of our days, and its day is almost the same as ours, being about 24 hours 37 minutes. Its surface gravity is only a little more than that of Mercury, or 37% of Earth's, and Mars's surface temperature ranges from around minus 255 degrees F (-153°C) to 70 degrees F (+20°C). It is the last of our terrestrial planets, with a thin atmosphere (mainly carbon dioxide with 2.7% nitrogen, 1.6% argon, and 0.7% being other gasses), although the atmosphere was much thicker in the distant past. Being so small, its central core cooled down relatively quickly and solidified. That meant that the planet lost its magnetic field about 4 billion years ago, allowing the solar wind to start blowing away the atmosphere. By about 1.5 billion years ago there was probably very little atmosphere left. Although Mars doesn't have an actual magnetic field, parts of its crust are still magnetic, and these areas have been seen to result in localized aurora.

It has seasons and ice caps that look somewhat similar to those on Earth, except that the icecaps are largely frozen carbon dioxide, not water ice, although ESA's Mars Express orbiter did find signs of several underground saltwater lakes beneath the south pole[cci]. It is also suggested that 99% of the water that once formed Mars's oceans is now locked away in minerals buried within the planet's crust, and has not been lost to space, as previously thought[ccii].

Like Earth, Mars was pummeled by asteroids during the period known as the Late Heavy Bombardment. Although it's a smaller target than Earth, it is also closer to the source of the bombardment, and there is some evidence that the bombardment of Mars continued up to about 4.45 billion years ago. So, if life did start on Mars, the planet probably wouldn't have cooled down sufficiently for life to get going until around 4.2 billion years ago[cciii]. Mars's chemically rich lakes and oceans may have been lost about 3.5 billion years ago, so that wouldn't have left much of an opportunity for life to evolve beyond simple single-celled creatures.

Although the planet has now cooled and largely solidified, it can't be described as geologically dead. NASA's InSight lander recorded about 500 marsquakes between February 2019 and September 2020, and two of the strongest were close to Cerberus Fossae, a young region that has undergone volcanism and other geological changes through to 2.5 million years[cciv] and with some activity even to the present day, as mentioned above.

Mars has two little moons, Phobos and Deimos, which may be captured asteroids or the result of the impact that formed the Borealis Basin. Phobos is spiraling slowly closer in towards Mars and will get torn apart by tidal forces before it impacts with the planet. Deimos is slowly moving away from Mars and might ultimately break free and have to be reclassified as an asteroid.

Outer Planets, the Gas Giants

The early astronomers only knew of seven planets or wanderers, and two of those were actually the Sun and the Moon, so there were only five planets as we now know them. Three of those are the terrestrial planets, which are smallish, but close to us, by which I mean "close" in an astronomical sense. The other two planets known to the ancients are a lot further away but are also a lot bigger. These are Jupiter and Saturn, the gas giants. The other two outer planets, Uranus and Neptune, were discovered more recently.

The outer planets are not rocky like the inner ones and are known as Gas Giants because they consist mainly of gasses, mostly hydrogen and helium. Actually, the outer two are sometimes known as Ice Giants because they have more ice crystals mixed in with the gasses.

The further we get away from the Sun, the colder it gets and, out around where the asteroid belt is, the Sun cannot heat matter above the melting point of ice, so there we have the Frost Line, and the gas giants are found out beyond that point. Their evolution probably started when ice particles began clumping together until they were massive enough for their gravity to begin pulling in the gas and dust from the primordial cloud that created the Sun and the solar system.

One puzzle about the ice giants (Uranus and Neptune) is the sparsity of ammonia observed, which is a lot less than that seen at Jupiter and Saturn. It is thought that the ammonia may clump together with water in these planets to form slushy hailstones, called mushballs, with a mass of around 2 pounds (1 kg) each, and that these get pulled down through the planet's atmosphere, hiding the ammonia from our view[ccv]. It is the methane in these two planets that gives them their bluish appearance, but why Uranus has a much paler blue than Neptune has been a bit of a puzzle. It is now thought that both planets have a haze in their atmosphere resulting from methane being broken down by the UV light from the Sun and then forming larger hydrocarbons, and the haze would look whitish at visible wavelengths. It would also be about twice as thick around Uranus, making the planet's intrinsic blue color look more "washed out"[ccvi]. These two planets also have complicated magnetic fields. There are quite a lot of mysteries with these planets, so it's time to send another probe out there to take a closer look and obtain some answers, and undoubtedly raise a lot more questions. It's good to see that a Uranus orbiter and probe is proposed for launching around 2032.

Jupiter

Jupiter is named after the Romans' principal god, with the name meaning Sky Father, and this was the Roman version of Zeus. Its symbol is supposed to represent the god's lightning bolt, but to me it looks like a 2 and a 4 rolled into one. It is the ruling planet of Sagittarius and Pisces, and it is also exalted in Cancer. The planet stands for growth, expansion, prosperity, and good fortune. It governs long-distance travel (which seems appropriate, seeing how far we have to travel to get to it), higher education, and the law (its gravity does help it to "police" the solar system), and it rules over Thursday.

Jupiter is the fifth planet from the Sun, about 87,000 miles (140,000 km) in diameter, which is over 10 times Earth's diameter, and its volume is about 1,300 times Earth's but it only has 318 Earth masses. It has 79 moons at last count, and it even has a faint ring, but that is nothing as impressive as Saturn's. Its most famous feature is the Great Red Spot, which is a giant storm about two or three times as big as the Earth and which may extend something like 300 miles (500 km) down into the gas giant. The Great Red Spot has existed for four centuries or more, at least as long as we have been able to see any detail on the planet. Well, we know that *a* spot has been seen for 400 years or so, but we can only be certain that we have been seeing the *same* Great Red Spot for about 150 years. That's still a lot longer than any storm I'd like to encounter.

Jupiter rotates faster than any of the other planets, taking just under 10 hours for one revolution, and it bulges noticeably at its equator because of its fast spin (29,000 mph or 46,600 km/hr). It is the largest of the planets in our solar system, with two and a half times the mass of all of our other planets combined. Despite its size, because it is so far away from us, it is not quite as bright as Venus in our night sky, but that still leaves it as the third brightest object we see among the stars. It orbits at around 490 million miles (800 million km) from the Sun, actually ranging from about 460 million to 508 million miles (740 million to 817 million km), making it about three and a half times as far out from the Sun as Mars is. The gas giants really are "outer planets" – they are way out there.

Its orbit of the Sun takes about 11.9 of our years, and the Earth overtakes it every 399 days, at which time it looks as if the planet has started moving backward for a time. That's an optical illusion caused by us viewing Jupiter from different angles against the background stars as we overtake it on the inside track. Its surface gravity (if it had a surface) is about 2.5 times that of Earth's, so you wouldn't jump very high there, and you'd get rather cold because it is about minus 162°F (-108°C).

It is called a gas giant planet because it consists mainly of hydrogen with some helium, and the planet has a kind of striped appearance. When the Galileo probe dived into Jupiter at the end of its mission, it found that hydrogen formed 90% of the gas. The wind got down to about 400 mph (650 km/hr.) and the temperature was about 300°F (150°C) before the probe vaporized around 95 miles (150 km) deep inside the planet. In a way, Jupiter could be called a failed star and it does produce heat inside it, but that heat is caused by the fact that gravity is still making the planet contract, not because of nuclear reactions like we find in the Sun. It probably has a rocky core with heavier elements.

Since it is largely gas anyway, the dividing line between the planet and its atmosphere is hard to pin down, but it does have a large atmosphere above the visible surface with winds that can be around 225 mph (360 km/hour), and with clouds of ammonia crystal. Its magnetic field is 20 times as strong as ours and can stretch for four million miles, and results in impressive aurora that also give off bursts of X-rays. These bursts have been found to coincide with vibrations in the planet's magnetic field, and it is thought that charged particles get caught in the waves and are smashed into the planet's atmosphere causing the X-ray pulses[ccvii]. If we could see Jupiter's magnetic field from Earth, it would look five times larger than the Moon in the night sky.

The planet's upper atmosphere is measured at a surprising 800°F (430°C), and this is believed to be due to its strong magnetic field and magnificent auroras. What is thought to be happening is that the charged particles driven by the magnetic field collide with the atmospheric particles, producing tremendous energy in the atmosphere around the poles, and the resulting heat then moves in waves from the poles towards the equator[ccviii].

Jupiter was the first planet to form from the cloud of gas, ice and dust that surrounded the infant Sun. The ice and dust started to collect together in clumps, maybe triggered by the shockwave from an exploding star within the star cluster where the Sun was born, and then gravity pulled some of these smaller clumps together to form the young Jupiter. As the ball of ice and dust grew, so did its gravity until it was strong enough to pull in and hold the gas from the cloud. That made it grow even quicker.

Around 4.5 billion years ago (September 3 in our "cosmic year" timescale), slowed by its passage through the gas cloud, Jupiter began spiraling in towards the Sun and disrupting the inner solar system. That probably caused a lot of the planetary collisions we were talking about with the terrestrial planets. It also cleared out a lot of the protoplanets and other matter around where Mars now orbits, robbing that planet of material and preventing it from growing bigger. That got Jupiter a name for being the bully of the solar system.

Saturn came to Jupiter's (and our) rescue, pulling it back out to where it now orbits. Had Saturn not interfered, Jupiter would most likely have continued spiraling in towards the Sun and disrupted the formation of Earth, Venus and Mercury. Consequently, we could have ended up with little in the way of any terrestrial planets, but we'd have the archetypical hot-Jupiter, orbiting maybe closer to the Sun than Mercury is now. Thankfully, that didn't happen and, as Jupiter moved back out, it affected the orbits of Uranus and Neptune too, and apparently made them switch positions, and it also may have thrown a fifth gas giant out beyond the Kuiper Belt.

Jupiter seems almost like a little solar system in itself, with so many moons surrounding it. Its four largest moons are called the Galilean moons and they are easily visible from even a fairly small telescope.

Io is the closest moon to the planet, and it is rocky and suffers from "land tides" due to the intense gravitational pull from Jupiter and the tugs from the other moons. That gives it a lot of tidal heating and makes it the most geologically active body in the solar system, with sulfur volcanoes all over the surface that seem to even deposit sulfur on the surface of Europa and some of the other moons. And, of course, it is constantly resurfacing itself with lava flows, so you won't find craters on Io because any that form are soon covered over.

Europa is rocky with an icy surface, and it gets stretched, squeezed, and heated by tidal forces resulting in it having a subsurface ocean that is believed to be about 60 miles (100 km) deep, hidden below the icy shell that can be up to around 18.5 miles (30 km thick). Also, the ocean has a rocky seafloor giving the reactions believed to be needed to bring about life. It seems that there are pockets of liquid water trapped in the thick ice shell of Europa, maybe 6 miles (10 km) or so wide, and only a kilometer or so below the surface. They may only last for tens of thousands of years, but there could be hundreds of such "lakes". They might develop as water from the ocean leaks into the ice crust or maybe arise from part of the crust melting. Some may be the source of plumes erupting from the surface. If there is life in the ocean, it could have become frozen into the shell and then be revived when remelted in a "lake" to start a new community[ccix]. Radiation from Jupiter's magnetic field makes Europa's ice-and-salt surface glow in the dark, and that glow varies depending on the ice-salt compositions across its surface[ccx]. Europa's core is believed to be iron-rich and to extend to a radius of about 370 miles (600 km), with a rocky mantle about 125 miles (200 km) thick on top of that.

Ganymede is a rocky-ice moon, larger than Mercury, and it also possibly has a subsurface ocean which may be partly due to tidal heating, but it is known to have a magnetic field which indicates that it still has an internal molten core that will be adding heat to the ocean. There are also signs that there may be cryovolcanic activity (ice volcanoes) resurfacing some of the icy moon.

Callisto is another rocky-icy body, but it orbits too far out from Jupiter to experience much in the way of tidal heating, even though its orbit is highly elliptical. However, it does have a subsurface ocean heated through radioactive decay. Its icy shell is expected to be 125 miles (200 km) thick, and it has a thin atmosphere of carbon dioxide.

Out beyond the so-called frost line (in other words, out where the gas giants are), we can find moons that are rocky like our Moon, and ones that are icy like comets, or a mixture of both. We also find moons that are big enough to pull themselves into a spherical shape, and others that are irregularly shaped. There is no specific size limit for what constitutes a moon, but let's say they could be anything from around the size of a double-decker bus on upwards.

Saturn

Saturn is named after the Roman god of agriculture, and its symbol, looking like a mixture of a t and an h, is supposed to represent the god's sickle. It is the ruling planet of Capricorn, and it is exalted in Libra. The planet stands for boundaries, practicality, reality, and structure, and it certainly shows an interesting structure with its rings. It governs ambition, career, authority, and hierarchy, and as you might guess it is associated with Saturday. You could say it showed its ambition when it started tugging mighty Jupiter around.

Saturn is another gas giant, and its fantastic ring system consists mainly of ice particles and some dust and rock. If the dinosaurs had built telescopes and studied Saturn, they probably wouldn't have seen the rings because the rings are a fairly new feature (as far as astronomical time periods go). It is believed that one or two of Saturn's icy moons spiraled in a bit too close to the planet and Saturn's gravity tore them apart, and their remains make up most of what we now see as the planet's rings. The rings are about 30 feet (10 meters) thick in places but span 170,000 miles (274,000 km) from the outer edge on one side of the planet to the outer edge on the other side. There are nine continuous rings and three arcs, and it's about 66,000 miles (106,000 km) across the width of the ring system itself (inner edge to outer edge). The interaction with some of Saturn's moons keeps the rings "corralled" and can create ridges higher than the Alps as the moons cause the rings to twist.

It is the sixth planet from the Sun, about 74,600 miles (120,000 km) in diameter (around nine and a half times Earth's diameter) although its rate of spin that gives it a ten-and-a-half-hour day makes it wider at the equator and flatter at the poles. It orbits at around a billion miles or, say, 1.5 thousand million km from the Sun, about twice as far out as Jupiter's orbit is, or ten times further out than Earth. Its orbital period is about 29.5 of our years. Its "surface" gravity is about the same as Earth's because, while it has a lot more volume, it is also a lot less dense. With its low density, it is said that you could float Saturn on the water in your bath, but it would leave a ring. Its surface temperature is about -218°F (-134°C), but its core is about 21,000 degrees F (11,700 degrees C). When we say "surface", it's the visible surface, not a solid one.

As a gas giant, it consists mainly of hydrogen, with some helium and other gasses, and it has been thought to have a core of metals, such as iron and nickel, surrounded by liquid metallic hydrogen which is then enclosed in a layer of liquid hydrogen, all under extreme pressure. However, an analysis of data from the Cassini orbiter suggests that Saturn has a fuzzy core with a sludge-like mix of ice and rock that extends out to about 60% of the planet's radius, containing a total mass equivalent to seventeen Earths[ccxi]. Its atmosphere has some water-ice clouds and ammonia crystals in its upper reaches which give the planet its yellowish hue, and powerful lightning has been seen there. It is also thought that lightning could turn some of the ammonia in its atmosphere into a carbon soot that hardens and compresses into graphite and then turn into diamond as it falls through the increasing pressure of the atmosphere. The planet has paler banding than that on Jupiter, and its storms are not as long lived as Jupiter's Great Red Spot. However, Great White Spots, that are giant storms, have been seen from Earth and seem to occur once every Saturnian year. But the most interesting weather phenomena is the hexagonal storm at Saturn's north pole, possibly caused by resonances creating standing waves. Saturn's winds are the second fastest in the solar system at about 1,000 mph (1,600 km/hr.) or more. It has a weak magnetic field, and auroral lights have been seen at the poles.

Saturn has 82 official moons but around 150 if you count the smaller objects outside of the rings. If we counted the particles that make up the rings as being moons (which, in a way, they are) we'd go from an amazing count to a totally ridiculous one.

Mimas is a moon of Saturn that might have a subterranean ocean, but it probably does not have a rocky seafloor, reducing the odds of finding life there. Mimas seemed to be too small to have an ocean, but it wobbles in its orbit in a manner that suggests that there is likely to be water moving around below the surface. And if Mimas has an ocean, then maybe other small icy bodies out there have them too[ccxii].

Hyperion was discovered in 1848, and Phoebe, which has a retrograde orbit, was discovered in 1899.

313 miles (504 km) diameter Enceladus, with a surface

area very similar in size to that of the nation Turkey, is believed to have liquid saltwater beneath its icy surface, and so is seen as a place where life might have started, and its salty plumes that shoot hundreds of miles into space at 1,400 mph (2,250 km/hr.) from near its south pole indicate strong water-rock interactions. Higher than expected volumes of hydrogen found in the plumes may indicate that there are hydrothermal vents on the seabed, similar to those found on Earth and which are oases of life in our ocean depths. However, there is some indication that the plumes might originate from inside the icy shell, not from the ocean itself. Those plumes also top up Saturn's E-ring. Enceladus's surface may be affected by the tides in the under-ice ocean, resulting in icequakes shaking the icy shell. Enceladus's icy shell is thought to be about 12.5 miles (20 km) thick in most places but thinning out to around 3 miles (5 km) near the south pole, and the ocean under it is expected to be around 25 miles (40 km) deep.

Saturn's largest moon, Titan, is rocky and icy and larger than Mercury. It is the only moon in the Solar System to retain a substantial, but very cold (-290°F or -180°C) atmosphere and it has methane clouds and rain leading to hydrocarbon rivers and lakes. Titan's largest sea, Kraken Mare, is estimated to be 1,000 feet (305 meters) deep. On top of that, or actually beneath it, there is also an under-ice liquid water ocean. Titan experiences seasons that may dry up the hydrocarbon surface lakes in one hemisphere at times, while raining down methane and ethane to fill up the lakes in the other hemisphere. In that way, the lakes may appear to migrate over a 29-year cycle.

Titan is moving away from Saturn at around half an inch (11 cm) per year, which is almost three times as fast as our Moon is moving away from Earth. Titan also affects Saturn's tilt, and it is estimated that if Titan continues its outward movement, the tilt of Saturn's axis of rotation could end up perpendicular to its orbital plane, so it would look as if it was rolling along its orbit just as Uranus does. That might happen in about a billion years from now, all things being equal, and if the movement continues then Titan's orbit would get disrupted and it would either be thrown out or it might crash back into Saturn. However, the Sun is likely to have started turning into a red giant before then and disrupted the orbits of all the planets, so all bets are off[ccxiii].

Rhea is the second largest moon of Saturn, and it seems to have a faint ring system of its own and a thin atmosphere.

Most of the other moons are small, such as walnut-shaped Pan that is about 20 miles (32 km) across and has created a gap in Saturn's rings. Most of the smaller moons weren't discovered until the planet was visited by the unmanned spacecraft named Cassini.

Those of Saturn's moons that are named are mostly called after the Titans of Greek mythology.

Uranus

Uranus is the only planet named after a character from Greek mythology, having the Latinized name of the Greek god of the sky, Ouranos. It is the ruling planet of Aquarius for some astrologers, and it is exalted in Scorpio. The planet is associated with genius, individuality and unconventional ideas, and its bluish color leads it to be linked with electricity. It governs societies, clubs, and groups dedicated to humanitarian and progressive ideals, and it rules radical ideas and revolutionary events, so I suspect governments may not like it very much. It is also a favorite for schoolboy jokes.

Uranus is not normally considered to be a planet visible to the naked eye, but on a really dark night, with a clear sky, no light pollution, and sharp eyes you might see it if you knew where to look. It is sometimes thought of as being a gas giant like Jupiter and Saturn, but it is now often called an ice giant because it has a different mix of materials from our two largest planets, having ices of water, ammonia, methane and some hydrocarbons along with the hydrogen and helium. The methane gives it its aquamarine or cyan (i.e., bluish) color because methane absorbs red light. Uranus seems to have no internal energy source, or very little.

Its core consists of ices and rock surrounded by layers of super-pressurized fluid water, ammonia, and methane, and its axis of rotation has been knocked sideways so that one pole points almost directly at the Sun at times, which makes the planet seem to roll along its orbital path for part of its orbit. Some of the rocks in the core might be a girl's best friend – diamonds. It is thought that the thick ice layers in the core could compress so much that they get hot enough for the carbon to turn into diamonds that would combine into "diamond bergs" floating in a layer of liquid carbon.

It is the seventh planet from the Sun, about 31,760 miles (51,120 km) in diameter (about 4 times Earth's diameter). It is the third largest of the planets, and orbits at around 1,865 million miles (3 thousand million km.) from the Sun, about 20 times further out than Earth is and twice as far out as Saturn. Its orbit takes about 84 of our years, but its day is about seventeen and a quarter hours with the planet revolving the opposite way to other planets (except for Venus). Its "surface" gravity is about nine tenths of Earth's, and it has the coldest planetary atmosphere in the solar system, getting down to minus 370 degrees F (minus 224 degrees C), while its surface is about -323°F (-197°C). So, the hottest planet and one of the coldest planet both spin in the opposite direction to the other planets. Its wind speeds can reach 450 mph (725 km/hr.), but it is fairly featureless, without the kind of bands that Jupiter and Saturn have. Uranus has a magnetic field that is tilted about 60 degrees from its axis of rotation, so everything about the planet seems to be a bit askew.

The planet has 27 known moons, and the named ones are called after characters in the works of Shakespeare and Alexander Pope, the five largest being Titania, Miranda, Umbriel, Oberon, and Ariel. It also has a dark ring system, with 13 known rings, and the rings were not positively identified until 1977. There is speculation that the presence of ammonia (which has been found on a number of icy bodies in the outer solar system) and radioactive decay in the core, might mean that moons such as Titania, Oberon, and Ariel have liquid under-ice oceans that might be about half a mile (one kilometer) or more deep[ccxiv].

Neptune

Neptune is named after the Roman god of the sea, and its symbol resembles his trident. It is a ruling planet of Pisces and is exalted in Leo. The planet is associated with creativity, idealism and compassion, but also things like deception (masquerading as a star, perhaps). It governs hospitals, prisons, and places of retreat from society like monasteries (it is certainly isolated, although Pluto visits occasionally).

Galileo is believed to have seen and sketched the position of Neptune, but it moves so slowly across the sky that it is thought that he just assumed it was a star, not a planet. It is the eighth planet from the Sun and is about 30,800 miles (49,500 km) in diameter, a bit smaller than Uranus. The really unique thing about this planet is that its position in the sky was calculated before it was seen, or at least seen while not being mistaken for a faint star.

It orbits at around 2.8 thousand million miles (4.5 thousand million km) from the Sun, which is one and a half times as far out as Uranus, 3 times as far as Saturn, and 30 times as far as Earth. Its orbit takes almost 165 of our years, while its day is about 18 hours (its rotation period against the back-ground stars is 16 hrs. 7 min). Its "surface" gravity is a bit more than Earth's and, being so far out, its atmosphere can get down to minus 360 degrees F (minus 218 degrees C). Its average temperature is -330°F (-201°C), while its core is about 9,000 degrees F (5,000 degrees C) resulting in the planet giving off two and a half times as much energy as it receives from the Sun. Its atmosphere has active and visible patterns, with the strongest winds in the solar system, 1,200 mph (2,000 km/hr.) or one and a half times the speed of sound, and an anticyclonic storm called the Great Dark Spot is seen, similar to Jupiter's Great Red Spot, although it seems to only last a few years at a time. As with all the outer giant planets, it has a magnetic field, and here it is tilted about 45 degrees from the axis of rotation.

We now know Neptune as the second of the Ice Giants, and it has 14 moons that we are aware of, plus a faint ring. Triton is its largest moon, which is believed to be a captured Kuiper Belt object because it has a retrograde orbit. It is larger than our Moon and it has a thin atmosphere of nitrogen and methane. Although Triton is one of the coldest places in the solar system, with temperatures down to around -390°F (-235°C), it is also geologically active with ice volcanoes sending frozen nitrogen five miles (8 km) high above the moon's south pole, and it could well be another moon with a subsurface ocean. It would be very appropriate to have Triton as a water world. Ice volcanoes have also built mounds in places and created smooth volcanic ice plains in others.

Pluto

Pluto is named after the god of the underworld and of wealth, while the dog in the Disney cartoons was named after the dwarf-planet, not the other way around. It is the ruling planet of Scorpio, and it is exalted in Virgo. The planet (okay, dwarf planet) is related to renewal and transformation (perfect for a body made largely of ices that go through heating and cooling as it follows its very eccentric orbit) and it is associated with power and personal mastery. It governs big business, wealth, mining, surgery and detective work (a lot of detective work has been taking place as the data from the New Horizons mission is examined), and it is associated with Tuesday, the same as Mars.

To many people, Pluto will always be considered a planet, but the astronomers do have a good reason for downgrading it, because it has been found that it is just one of a group of similar objects, and not even the biggest of them. Eris is the largest known Kuiper Belt object and there could easily be bigger ones yet to be discovered. But Pluto is still fascinating, and many people think that it should keep its 'planet' designation. Pluto does meet the modern requirement for being a planet in regard to being spherical and orbiting the Sun, but it doesn't meet the requirement regarding "dominating" its orbit. In fact, its orbit sometimes crosses Neptune's orbit, with the result that for part of the time when it was considered a planet, it was the eighth planet from the Sun, not the nineth.

A large part of its surface area is covered in a reddish material that was thought might be tholins (a sort of mud) but testing that idea and seeing how the radiation would affect tholins might rule that idea out. Another idea is that the texture of the surface causes particular light absorption and thereby creating the color[ccxv].

It is about 1,480 miles (2,380 km) in diameter, less than one fifth of Earth's width and about two-thirds that of the

Moon, but it is over twice the diameter of the largest asteroid. Its orbit ranges from about 30 times as far away from the Sun as we are to almost 50 times and takes about 248 of our years to complete one circuit. Its day is almost six and a half Earth-days but in a retrograde direction. Its surface gravity is about a fifteenth that of Earth, so it's the place to go if you want to lose weight. Pluto's surface temperature is about -387°F (-233°C), so you might want to bundle-up while visiting it!

It consists of rock and ice and has about half the density that the Moon has. Surprisingly, it does have a bit of an atmosphere, consisting of a mixture of nitrogen, methane and carbon monoxide gases. Sputnik Plenum looks like a large smooth icy glacier and the planet has cryovolcanoes such as Wright Mons and Piccard Mons where water or soft ice seeps out[ccxvi]. They are like shield volcanoes 100 miles (160 km) wide and 3 miles (5 km) high, but it's not molten rock that forms them but molten ice (i.e., water) spewing out from inside the planet and then freezing again. Pluto also has vast basins filled with frozen nitrogen that is expected to have a texture like soft ice-cream. The lack of craters on Pluto shows that it is geologically active, resurfacing itself regularly.

Our first (and, so far, only) spacecraft to visit it, New Horizons, has sent back some fascinating images and data that will keep scientists busy for years. Pluto was still officially a planet when New Horizons launched, but not when it arrived at the planet, I mean dwarf-planet. They expected to find a dead frozen world, but instead found a geologically active place that likely has liquid water, maybe an ocean below the surface and could be home to life. The liquid water ocean, with a dash of ammonia in it, is believed to exist below the area known as Sputnik Planitia, beneath about 3.7 miles (6 km) of ice and kept liquid due to heat from radioactive decay.

Even though it is now known as a dwarf-planet, Pluto has 5 known moons, the largest by far being Charon which is 753 miles (1,212 km) in diameter with a fractured surface that probably came about as the moon cooled and its center froze. Its red-colored polar cap is probably the result of tholins that made their way from Pluto somehow (or whatever it is that looks red on Pluto). The four small moons are Hydra, Styx, Kerberos, and Nix, and they are about 25 miles (40 km) across and spinning crazily.

Planet 9

Astronomers have discovered indications of a ninth planet, far out, about 93 billion miles (155 billion km) away, or nearly one light-week, that appears to be disrupting distant Kuiper Belt objects. If it really is a planet out there, it would be about ten times the mass of Earth, and there are various ideas about what it might be.

Firstly, it could be a rocky super-Earth. Being large, it would retain the heat in its core for a long time, so it could be covered with active volcanoes. Super-Earths are a common type of exoplanet although our solar system lacks any, that we know of. Maybe one or more formed and either got thrown into the Sun or out of the solar system when Jupiter disrupted the solar system in its formative years.

This could be such a planet that didn't get completely thrown out.

Secondly, it might be like a smaller version of Neptune. Models of how the solar system was born seem to indicate that there should have been five gas giants, not the four we see now. Maybe this one got thrown out into a distant elliptical orbit when the gas giants were jostling around. Being so far out and very cold, its atmosphere could be nearly transparent making it appear like a massive translucent bluish ball lit up by intense thunderstorms within it.

A third possibility is that it is a planet that we captured from another solar system. When the Sun was born it would have been close to other newborn stars in its "stellar nursery" gas cloud and might have captured a planet tossed out of another solar system.

As a fourth option, there is a slight chance that Planet 9 (if it exists) could be a primordial black hole[ccxvii] (assuming that primordial black holes exist). As such, it might be about the size of a grapefruit, making it very difficult to see at that distance.

But maybe, just maybe, astronomers might have seen Planet 9, except it doesn't seem quite right. Seeing Planet 9 in optical light is highly unlikely at that distance, but it might have a heat signature that could be picked up with an infrared telescope. A study of the data from IRAS (the Infrared Astronomical Satellite), that was launched in 1983 and operated for nine months, gave a potential sighting of Planet 9 very close in the sky to where it was calculated to be. The problem is, it might not have been a planet that was being seen and, if it was, it is estimated to be only about

3 to 5 Earth-masses, which seems to be too small, and its potential orbit appears to be a bit different from what is expected too. It's an intriguing find though[ccxviii].

Asteroids, Comets and Meteors

The asteroid belt (or minor planet belt) lies between Mars and Jupiter, and mostly contains objects composed of rock and metal like the terrestrial planets. Some asteroids orbit outside of the main belt and can cross the orbits of the inner planets.

A survey carried out using images from the Sloan Digital Sky Survey (SDSS) identified over one and a half million asteroids, about a third of which had not been recorded previously[ccxix]. When you limit the size to 0.6 miles (1 km) diameter or above, there are around 822,000 known asteroids out of about 1.5 million asteroids estimated. However, there's no official size limit and if you limit the size of what you call an asteroid to 100 meters or so across, there are believed to be over 150 million of them. Even so, the total mass of the asteroids is probably less than that of the Moon.

Ceres is the largest asteroid, about 530 miles wide and it contains about 30% of all matter in the asteroid belt. There are also indications that it may have a subsurface ocean that is finding its way to the surface and leaving the whitish salt patches seen on the surface. It is thought that Ceres might have started its life further out, maybe in the Kuiper Belt, and got drawn in when the gas giants were doing their dance. Vesta is another asteroid that is sometimes referred to as a dwarf planet, but at about 329 miles wide, and with a not really spherical shape, it is more correctly called a giant asteroid.

The 1000th asteroid was discovered in August 1923, so most of them have been discovered in this past century. Almost all of them appear to orbit within the band defined by the zodiacal constellations, i.e., they are close to the ecliptic[ccxx]. In science fiction movies, asteroids are normally depicted as orbiting very close to each other, with spaceships having to maneuver cautiously between them. However, in reality, if you were standing on an asteroid, it is unlikely that you would be able to spot another in the sky. That doesn't mean that collisions can't happen and haven't happened in the past. Indeed, we can see the results of such collisions in families of objects that share similar orbits[ccxxi]. The Baptistina Family is a collection of objects in the asteroid belt that is estimated to have resulted from the destruction of an object about 110 miles (170 km) across. It was thought that the object about the size of Mt. Everest that finished off the dinosaurs might have been part of the Baptistina Family originally, but that has been disproven. However, the Chelyabinsk meteorite might have been from that Family[ccxxii].

Jupiter has captured some asteroids (maybe around 10,000 rocks of various sizes) which are known as the Trojan asteroids, and they occupy the places ahead of and behind Jupiter where the gravity of Jupiter and the Sun balance out (i.e., two of its Lagrange points). At least one of the Trojans in the leading bunch has a moon of its own.

The Trojans are classified into three groups based on their color, which probably indicates different compositions and may indicate that they were initially formed in different regions of the solar system[ccxxiii].

Just to get an idea of how vast space is within our solar system, we can take a look at the Lucy mission currently underway to explore the Trojan asteroids that are in Jupiter's orbit. When do you think the Lucy spacecraft will be at its closest point to Jupiter? It turns out that it will be when the spacecraft passes Earth to get a gravity assist to send it on its way to the asteroids[ccxxiv].

Other planets can pull asteroids or similar bodies into their orbits too. Earth is known to have five trojans (some are more trojan-like since they don't actually orbit Lagrange points), and one of them seems to have an unexpected origin. An investigation into the reddish color of Kamo'oalewa seems to indicate that it is an object that was once part of the Moon but got blasted off by an impact and ended up in the same orbit around the Sun as Earth[ccxxv].

Earth's two real trojan (with a small t) asteroids orbit at the L_4 point, the leading Lagrange point. 2010 TK_7 is about 980 feet (300 meters) diameter, and 2020 XL_5 is around 1,150 feet (350 meters) diameter[ccxxvi]. 2020 XL_5's wobbly orbit takes it about 10% closer to Venus's orbit at times than Earth is, and this trojan could be dislodged from its orbit in about 6,000 years and become a regular NEO (near Earth object)[ccxxvii]. Our trojan and trojan-like companions are fairly well behaved, but some other asteroids that are known as NEOs are viewed as potential threats to Earth, as the dinosaurs could confirm if they were still around.

It doesn't take an asteroid as big as the Chicxulub one to cause substantial damage. A meteorite struck in the Bolavan Plateau in southern Laos, creating a 10.5 x 8-mile (17 km x 13 km) crater that is now covered by an ancient lava flow. The roughly 1.25 mile (2 km) wide impactor would have sent pillow-sized boulders at supersonic speeds (say 1500 ft/sec or 460 meters/sec) into the air, and everything within 300 miles (500 km) would have been incinerated. Then hailstorms of black glassy debris (tektites) would have started raining back down over 10% of the Earth's surface and we find such artifacts from Asia to Antarctica[ccxxviii]. Also, the midair explosion of a large meteorite over Tunguska in Siberia in 1908 flattened 1,000 square miles (2,600 square km.) of forest. Happily, that was in a basically uninhabited region[ccxxix]. Asteroid Day is a day dedicated to preparing for any future potential asteroid impact, and it is observed annually on June 30, the anniversary of the Tunguska blast.

Apophis is one NEO that people are keeping a particularly close eye on, but observations indicate that Apophis will not impact Earth for at least 100 years[ccxxx]. Of course, that's still real soon in astronomical terms, but we've no evidence yet that it is set on an actual collision course.

The Centaurs are asteroids or comet-like objects that orbit within the realm of the outer planets, and include Chiron, Asbolus, 1994 TA, Nessus, Chariklo, and Pholus. Chiron was discovered in 1977, and it was initially called an asteroid but then became known as the first discovered Centaur. It is believed that it was "recently" dislodged from the Kuiper Belt, as were the other Centaurs, but they cannot last long within the realm of the giant planets because they either collide with a planet or maybe get thrown from the solar system. Although there's a lot of empty space out there, the giant planets have a long reach with their gravitational pull. The two biggest of the Centaurs are Chiron and Chariklo, which are 137- and 205-mile diameter (220 and 330 km)[ccxxxi] respectively. Some of the Centaurs are active, sporting comae (fuzzy, comet-like atmospheres, and even jets of gas and dust), and their colors range from neutral gray to extremely red with surface compositions showing water ice and organic compounds. A few have moons and some, including the two biggest, have rings. Chiron has also been seen to have carbon monoxide ice that is sublimating into space (melting directly into a gaseous vapor). Chiron is smaller than Pluto (1/10 the size, one thousandth the mass) and larger than Arrokoth (10 times the size, and 1,000 times the mass). Representing the smaller-sized Centaurs, we have 29a/Schwassmann-Wachmann 1 (SW1), that is only 38 miles (60 km) in diameter.

Beyond Neptune's orbit lies the Kuiper belt which contains swarms of planetesimals and dwarf planets, like the asteroid belt, but they are more likely to be icy bodies, rather than solid rock. When they get pulled out of their orbit, they can become short-term comets, that is, comets that return fairly regularly. Kuiper Belt dwarf planets include Pluto, Eris, Sedna, Makemake, and Haumea, but most of the Kuiper Belt objects are irregularly shaped smaller objects like the "snowman" shaped Arrokoth that New Horizons visited.

New Horizons also managed to image two other Kuiper Belt objects, 2011 JY31 and 2014 OS393, both of which are actually pairs of objects in tight orbits around each other, or rather around their joint centers of gravity. The objects making up 2011 JY31 are both about 30 miles (50 km) across and are separated by about 125 miles (200 km), while the 2014 OS393 objects are about 20 miles (30 km) across and separated by about 95 miles (150 km)[ccxxxii]. Interestingly, Arrokoth could be described as a pair of bodies that were touching and merged, so maybe some of the collisions that built up the planets were not always the head-on smash-ups that we imagine.

Another Kuiper Belt object, 2018 AG37, nicknamed Farfarout, has been confirmed as the most distant solar system object observed, being 132 times the distance from the Earth to the Sun which is almost four times further from the Sun than Pluto is. It is 250 miles (400 km) wide[ccxxxiii].

Long-term comets are believed to come from the Oort cloud, a collection of trillions of icy objects believed to orbit about one to two light-years out from the Sun, about halfway to Proxima Centauri. It is thought that some Oort Cloud objects may switch from orbiting our Sun to orbiting the Alpha Centauri system and back again.

There is a comet named Bernardinelli-Bernstein (C/2014 UN271) that is the largest known visitor from the Oort Cloud and is about twice the size of the next biggest comet known to have come our way, Hale-Bopp. It's calculated as being about 85 miles (137 km) across[ccxxxiv] and is believed to have come from about 40,000 AU away in the

Oort Cloud, or around 15% of the distance to Proxima Centauri. By 2021 it had reached a distance from the Sun that was similar to the orbit of Uranus, and it will make its closest encounter with the Sun in 2031, but even then, it will still be beyond the distance of Saturn's orbit.

As the icy body of a comet on its inward journey from the Oort Cloud or Kuiper Belt gets to around 500 million miles (800 million km) from the Sun, the heat starts to vaporize its surface and it begins to look like a fuzzy ball and form a tail that points away from the Sun. The closer the comet gets to the Sun, the stronger the solar wind becomes, and that wind scours molecules from the comet's surface, forming a second tail. The tails can be tens of millions of miles long[ccxxxv].

If you see a shooting star, it is actually a meteor, a small lump of rock maybe about the size of a grain of sand that is burning up as it enters the atmosphere, and if it survives to reach the surface of Earth before it is fully burned up, its remains are known as a meteorite. It is estimated that around 5,000 tons (4,500 Megagrams) of space dust falls on Earth annually.

Generally speaking, meteor showers are the result of the Earth passing through the orbit of a comet, and the meteors are little grains of matter that had been ejected from the comet on earlier close encounters with the Sun. The one exception to this is the Geminid meteor shower, and that is the result of debris from an asteroid, not a comet. The asteroid in question is named 3200 Phaethon, and on its closest passes to the Sun it is only at a distance of 0.14 astronomical units from the Sun, which results in the Sun heating its surface to around 1,390°F (750°C). At those temperatures, it is believed that sodium close to the surface of the asteroid vaporizes and expels other surface rock into space. That expelled matter becomes the meteors that illuminate our pre-dawn sky during December.

Not all of the smaller objects we see passing through the solar system are actually part of our system. Recently we have seen 'Oumuamua (1I/2017 U1) which was detected October 2017 and 2I/Borisov that was seen in August 2019, and both of them arrived from outside the solar system. 'Oumuamua ("scout" or "messenger" in Hawaiian) was cigar-shaped, six times as long as it was wide, and it was moving slower than Borisov did. It was brightening and dimming by a factor of 12 every 4 hours, indicating that it was apparently tumbling. It seemed to be rocky, in that it had no cometary coma, but it accelerated, probably from something like cometary outgassing. Some people suggested that it was an alien spaceship or a solar sail. It has also been suggested that it was a "hydrogen iceberg" of interstellar molecular hydrogen ice (H_2). Such H_2 has been proposed as a candidate for dark matter. If it was such a "hydrogen iceberg", that might account for acceleration through outgassing but showing no coma.

Another idea is that it might have been solid nitrogen, blasted off the surface of a planet like Pluto in an alien solar system maybe 500 million years ago, then travelling through interstellar space while being slowly eroded by cosmic rays. The approximately 755 feet x 115 feet x 115 feet (230 meters x 35 meters x 35 meters) object entered the solar system around 1995, made its closest approach to the Sun as it swung around it on September 9, 2017, had its closest approach to Earth around October/November 2017, and is expected to exit the solar system around the year 2040[ccxxxvi].

Borisov is an alien comet, and it was sublimating CO (carbon monoxide) indicating that it had been cold for a very long time because CO sublimates at very low temperatures. It has more carbon monoxide than water, completely unlike our comets[ccxxxvii].

Explosive Creation of the Universe's Ingredients

Soon after the Big Bang, the universe had created hydrogen, a bit of helium, and miniscule amounts of some other light elements. Then, clouds of the gas collapsed, and fusion started, resulting in more helium being created. As the hydrogen in a star's core gets used up, the helium and then even heavier elements start fusing, producing carbon, nitrogen, oxygen, and finally iron (if the star is big enough) that can be turned into the steel that is used to hold our buildings up, among other things. However, iron will not hold up a star, and in fact it is deadly to them because it will cause a giant star to collapse and explode as a supernova, but that blasts the heavy elements out and they can be made available for future stars and planets.

The copper that is used in the wiring of your house comes from the next generation of stars that began with some of the heavier elements already peppered in them. As these smaller stars begin expanding and cooling into red

giants, they form heavier elements, including copper, by atoms exchanging neutrons which then decay into protons
and thereby create newer, bigger atoms. Gold is the element that the alchemists were mainly interested in and is
one that has been the foundation of many economies, and
that comes from the greatest explosions since the Big
Bang. It takes something like two neutron stars spiraling
around each other until they coalesce in a neutron star merger, releasing about as much energy as the Sun releases in
its lifetime, but in a single second.

The oil that we have been using as a fuel, and are hopefully weaning ourselves off of, comes from the remains of
forests and dinosaurs, from long ago, that got buried after
a comet or asteroid crashed into Earth. That mix got converted to oil by the intense pressures underground caused
by Earth's gravity. With all these explosions and catastrophic events, it seems that the universe is out to kill us,
but actually it's the way the universe grows. Generally
speaking, everything we use and everything we are made
of, comes from some sort of explosion in space or on
Earth. Okay, the hydrogen came about as the universe
cooled off after the Big Bang, but that was the biggest explosion ever and created space, rather than occurring *within*
space.

Chapter 10: Earth, Moon and Water

Earth's Composition

Every planet formed from the same ingredients, namely the gas and dust that surrounded our young Sun. The terrestrial planets were in a warmer region, so elements that turn to gas at low temperatures are rare in that area because they got blown outward by the radiation from the Sun. Consequently, the terrestrial planets are largely rocky places with atmospheres and surface water that came later (released through geologic processes or delivered by impacts) but the planets ended up very different from each other due to their differing situations. All terrestrial planets are, or have been, subject to a greater or lesser extent to things such as cratering, sedimentation, landslides, and volcanism which have remade the surface and atmosphere[ccxxxviii].

Our blue watery planet Earth, also known as Terra, is the third planet from the Sun, and the first one that has a moon. Earth has a diameter of 7,918 miles (12,742 km), a mass of 1,317 x 10^{25} pounds (5.972 x 10^{24} kg), and an atmosphere only about 60 miles (100 km) thick that now consists of 78% nitrogen, 21% oxygen, 0.9% argon, and 0.1% other gasses such as carbon dioxide, water vapor, etc. But Earth didn't start out with that atmospheric mix. The atmosphere has been completely remade twice, firstly by volcanic activity and meteorite impacts, and then by the emergence of life.

When you create a planet by smashing planetesimals together, you generate a tremendous amount of energy in the form of heat, and that can effectively leave the planet in a molten state. In such a condition, the different materials can start to separate out, with denser materials such as iron sinking to the center, and the lighter rocks floating to the surface to create the planet's crust[ccxxxix]. We see that kind of thing with the Earth's solid inner core surrounded by the molten nickel-iron outer core that generates our protective magnetic shield. That field reverses its polarity approximately every 700,000 years, but we're not really sure why or how. The temperature of Earth's inner core is still about 9,400°F (5,200°C) which is not much different than the "surface" of the Sun. Outside of the core we have the soft-ish mantle that can bubble up through the outer surface crust in places due to geologic seismic activity and volcanism. What nobody was expecting to find was two massive regions situated between the core and the mantle that are hotter than the mantle and, with one under the Pacific and the other under Africa, they constitute almost 10% of the mantle and show up when we measure seismic waves traveling through the Earth. Sometimes called superplumes, they are active and have been associated with various seismic phenomena including some island building and the movement of continental masses. Ideas for what caused these superplumes include the breakup of the once great continent of Pangaea to the remains of the Mars-sized planet Theia that is believed to have impacted Earth and resulted in our Moon being formed[ccxl].

Our homeworld has a water cycle, whereas Mars now has a carbon dioxide cycle, but water and carbon dioxide behave differently. Earth is unique and bizarre, the only one that still has abundant liquid water on its surface driving an active water cycle, and the only one known to still have active plate tectonics creating and destroying distinct pieces of crust and releasing volatiles from the interior that help create and maintain our atmosphere. We also have our oversized Moon that stabilizes our 23.4° planetary tilt giving us regular seasons, without which the climate could destabilize in just a few thousand years, even without our help.

Life has played a major role in shaping the planet. For instance, vegetation has been the main player in providing us with our current air containing 21% oxygen. Earth started with an atmosphere rich in methane and carbon dioxide, but photosynthesis converts sunlight and carbon dioxide into energy and, thereby, releasing oxygen. There was already life here before photosynthesis started but then, around 2.4 billion years ago (October 29 on our "cosmic year" scale), things changed and nearly all of the early lifeforms were killed off by rising oxygen levels after photosynthesis arrived on the scene. However, that paved the way for life as we now know it. Earth has a huge number of minerals, and the big explosion in mineral diversity occurred around 600 million years ago (December 16 in our "cosmic year"), coinciding with the emergence of life on land, showing that life and geology coevolved.

Recently, humans have been the dominant force for change on the planet, removing and replacing vegetation, exhausting and rerouting water supplies, populating and reshaping coastlines, not to mention the greenhouse gasses we've been pumping out.

Venus and Earth seem to have started out much the same, with Earth being only 1.1 times wider and 1.2 times as massive as Venus. Both were made of roughly equal amounts of the same material, and Venus supposedly once had similar amounts of water, but it's changed. Now Venus rotates retrograde and, as far as we know, it has no plate tectonics even though its core is still hot, and it has an oppressively thick and hot atmosphere. Let's try not to follow in Venus's footsteps until the Sun forces that situation on us.

Earth's Motion Through Space

Earth is spinning on its axis at about 1,000 mph (460 meters per second) at the equator and is orbiting the Sun at about 67,000 mph (108,000 km/hr.). During the morning twilight you will be on the side of the Earth facing forward as the planet moves in its orbit, and at the evening twilight you will be facing back along the path we just traveled. The Sun is carrying us around the Milky Way at about 560,000 mph (900,000 km/hr.) while the Milky Way galaxy is being pulled towards the so-called "Great Attractor" at 1,340,000 mph (2,160,000 km/hr.). Analysis of the CMB data seems to indicate that our Local Group of galaxies is moving at around 1,400,000 mph (2,260,000 km/hr.) in relation to that background radiation, but individually we are moving at 823,000 mph (1,325,000 km/hr.) with respect to the CMB. So, we are traveling in a cavorting spiral through the universe, like a crazy helter-skelter ride. Feeling dizzy?

The Moon

The Moon has sometimes been associated with the Roman goddess Diana, it is the ruling "planet" of Cancer, and it is also exalted in Taurus. It is associated with unconscious habits, memories, moods, and the ability to react and adapt to those around us. It rules over Monday, or Moon's-Day. The Moon is the only one of the ancient astronomers' "planets" (with the exception of the Sun itself) that doesn't revolve around the Sun, because it revolves around Earth (which does revolve around the Sun, of course).

Earth's only natural satellite, also known as Luna, is believed to have formed from a collision of a Mars-sized planet, that we call Theia, with the early Earth. It was apparently more of a glancing blow than a head-on pile-up, but was still violent enough to destroy Theia, melt the surface of Earth, and send large amounts of both into orbit.

Simulations of the 30 minutes after the Theia impact show a massive chunk of Earth being gouged out and the debris blasted into space. Much of the fiery debris then came raining back down onto Earth, some of it being huge fragments that had traveled halfway around the globe, and within 48 hours the remainder had formed a disk around Earth that would eventually recombine to form the Moon[ccxli]. Along with a sizable portion of Earth's crust, most of Theia's crust ended up in the Moon, although its iron core probably sank into Earth's core.

One problem that has been pointed out about the Theia idea is that ratios of oxygen-16 to oxygen-18 are very similar on the Earth and the Moon, but different from elsewhere in the solar system. The one exception to that is meteorites called enstatite chondrites, and if both Earth and Theia had been built from them, there's no problem[ccxlii]. Other studies suggest that most of the debris blasted into orbit came from Earth's crust, rather than from Theia, which would explain the similarity of isotopes between Earth and the Moon. The simulations also indicate that the Moon might have started coming together within hours of the collision[ccxliii].

The Moon is known for its very old, heavily cratered surface and its volcanic maria or seas that are really large impact basins that were flooded and filled in with dark lava, not water. It has many features to observe, including craters, mountains, rills, scarps, ejecta rays, and more. Its surface is actually fairly dark but reflecting the Sun's rays makes it the second brightest object in the sky and the brightest in the night sky, which is why it has been such a fascination for people down through the ages.

It is about 2,160 miles (3,480 km) in diameter (a bit over a quarter of Earth's diameter), and it has almost exactly the same apparent size in the sky as the Sun, which is great for creating total solar eclipses. It orbits at around 250,000 miles (400,000 km) from Earth, although it is moving away from us at about 1.5 inches (4 cm) a year. The distance between the Earth and Moon is such that you could squeeze all the planets in the solar system between them, although I'm not suggesting that you try. It takes about 27.3 days for the Moon to go around Earth once, while it spins on its axis in the same time period, so the same side of the Moon always faces us. When we take into account Earth's motion around the Sun, we get a lunar month – New Moon to New Moon – of 29.53 days. Its surface gravity is about one-sixth of Earth's, and the Moon's surface temperature can get down to minus 396 degrees F (minus 238 degrees C) in some areas where the Sun doesn't reach.

Some areas might never even get reflected sunlight, and it is thought that they might have temperatures getting down to -420°F (-250°C), making them some of the coldest places in the solar system[ccxliv].

It is now the largest moon relative to the size of its planet. Charon (Pluto's largest moon) is larger in comparison to Pluto than the Moon is to Earth, but Pluto is no longer officially a planet, so hard luck Charon, you've lost your status too.

Magmatic or igneous rocks have been found on the Moon that are less than 2 billion years old in Oceanus Procellarum, which is showing that volcanic activity was going on considerably longer than had been expected. One possibility is that radioactive decay kept some regions of the Moon hotter for longer, or it could be that the Moon got more tidal heating when it was orbiting closer to Earth[ccxlv], which would imply a more elliptical orbit back then. Although the Moon's "oceans" consist of magma, not water, there appears to be ice in some regions of the Moon where the Sun's light never reaches[ccxlvi].

The Moon has a faint magnetic field, about a thousandth the strength of Earth's, and how that is caused is still a matter of speculation. There are indications from some locations of a strong magnetic field in the past, but it is believed that these may have been caused by asteroid impacts magnetizing the moon rocks[ccxlvii].

The Moon has influenced our culture substantially and shows its influence on the Earth twice daily as the high tides hit the shoreline. Earth and Moon pull on each other, creating tidal bulges and, due to Earth's rotation, our tidal bulge will always be just ahead of the Moon, tugging and accelerating the Moon slightly and enlarging its orbit.

Precession of the Equinox

Gravitational tidal forces from the Sun and Moon tug on the spinning Earth and cause the axis of rotation to change slowly. This is somewhat akin to how the tug of Earth's gravity causes the vertical axis of a spinning top to rotate slowly in Earth's gravitational field. As a result, the North Celestial Pole rotates anticlockwise around the North Ecliptic Pole as viewed from Earth. For the southern Celestial Pole, the motion is clockwise. As the Celestial Poles move relative to the Ecliptic Poles, the Celestial Equator moves relative to the Ecliptic. The motion is such that the Equinoxes move westward along the Ecliptic, with a complete circle taking 25,800 years, which is about 1° of movement in relation to the "fixed" background stars approximately every 72 years. With the zodiacal "houses" each covering 30° along the ecliptic, that means that about 2,150 years constitutes the "Age" of a zodiacal "house". This "Precession of the Equinoxes" was first noticed by the ancient Greeks.

The Short Story of Water

Soon after the Big Bang 13.8 billion years ago, the energy cooled enough for hydrogen to form, and that hydrogen later collected together, and gravity collapsed the hydrogen clouds into the first stars. That was probably about

100 million years after the Big Bang (January 3 in the "cosmic year"). The nuclear fusion at the heart of these giant stars created new elements, including oxygen, and when those early stars exploded as supernovae, these new elements got blasted out into the gas clouds within the islands of stars known as galaxies. The atoms in the clouds now included lots of hydrogen and plenty of oxygen, and it only takes two hydrogen atoms and one oxygen atom to come together and you've got a small drop of water. So, the universe soon had plenty of water among the clouds of gas and dust, with that water normally being more like ice because baby it's cold outside.

4.6 billion years ago (September 1 on our "cosmic year" calendar), a supernova (or something like it) sent shockwaves into a cloud of gas and dust in the young Milky Way galaxy, and that pushed the gas together and a star cluster started to form, which included our Sun. Any water that got drawn into the young Sun soon got superheated and ejected as steam, and most of the water in the gas relatively close to the Sun got broken up and blasted away into the further reaches of the solar system by the solar wind. But out there, it was cold enough for the water to become ice and start to clump together with the dust until the clumps grew big enough for gravity to start pulling in more gas, dust, and ice.

Soon after that, mighty Jupiter was born, about 4.5 billion years ago (September 3 in our "cosmic year"). But the gas, dust and ice had also formed millions of smaller asteroid-like or comet-like objects, and when the drag of the gas cloud started to slow Jupiter's orbital speed and make it start to spiral closer to the Sun, it pushed or pulled a lot of those objects into the inner solar system. Some of those asteroid and comet-like objects combined with the other matter and protoplanets in the inner solar system to form Mercury, Venus, Earth, Mars and at least one other Mars-sized planet that we call Theia. But Theia collided with the young Earth in a glancing blow, blasting off much of the Earth's crust and sending it into orbit along with some of Theia and formed the Moon. Theia's metallic core got drawn down to combine with Earth's core. Earth's surface was like a sea of molten lava, and any water it had at that time near the surface would have been vaporized, but there was still plenty of water trapped inside Earth. Some of that trapped water got blasted out by the many active volcanoes over time, and as the surface of the Earth cooled, the first rivers, seas, and then oceans appeared.

Out beyond Jupiter, the other gas giants had formed. There was Saturn, Neptune, and further out there was Uranus, and as Saturn started to spiral in towards the Sun it got into resonance with Jupiter. For every two times that Jupiter orbited the Sun, Saturn did once, and their constant gravitational interactions started to pull Jupiter outwards again, and Saturn along with it. That disrupted the orbits of the other gas giants, throwing Neptune out beyond Uranus which in turn disrupted the Kuiper Belt. The result was a rain of comets falling into the inner solar system, with maybe one comet hitting Earth per month for about 100,000 years. That period is known as the Late Heavy Bombardment and started about 4 billion years ago and lasted through to about 3.8 billion years ago (September 14 through 22 on our "cosmic year" calendar). The icy comets added to the amount of water in our oceans, and as things calmed down, life was able to take off, leading to us.

Mercury was too small to gravitationally hold onto any water it had, especially being that close to the Sun, although it seems to have some ice in sheltered spots. Venus apparently once had oceans but being that much closer to the Sun it was a lot hotter and the resulting runaway greenhouse effect boiled off the water, and it got broken up and blown away by the solar wind. Mars also had oceans, but being so much smaller than Earth, it cooled more quickly and lost the magnetic field that protected it from the Sun's radiation. Consequently, it too lost its surface water, although it still has water trapped beneath the surface, and sometimes that water even seems to break through today. That left Earth as unique in the solar system with liquid water in abundance on the surface.

Chapter 11: Exoplanets

Democritus, back around 400 BCE, spoke of alien planets of various sizes, some with moons, some without moons, some orbiting a star, some orbiting a pair of stars or maybe more. We are now discovering that worlds like this really exist.

The first confirmed exoplanet (orbiting a pulsar of all things!) was discovered in 1992. That was the system known as PSR B1257+12, and three exoplanets have been found there, but they are constantly bathed in radiation that would be deadly for any conceivable lifeforms. These three nearly Earth-sized planets either somehow survived the supernova explosion that formed the pulsar, or, more likely, grew from the remains left over after the blast. They are orbiting closer to the pulsar than Mercury is to the Sun, so they are being seriously irradiated.

The first known planet orbiting a Sun-like star was found in 1995, and by August 2021 we had discovered nearly 5,000 exoplanets in almost 3,700 different systems, including ones with two or three "suns", planets that are basically gigantic solid diamonds, and at least one that should be raining glass. There are also Earth-like planets in stable orbits around Sun-like stars in the Goldilocks zone where liquid water can exist, although out of the exoplanets found by mid-2020, only about 50 of them were considered (optimistically) as being habitable, i.e., they could have liquid water and a rocky surface, but only about twenty of them had Earth-like temperatures[ccxlviii]. TESS (the Transiting Exoplanet Survey Satellite) was launched in 2018 and had itself found over 5,000 potential exoplanets by March 2022, and it was still going strong.

The nearest star system we know of that contains at least one exoplanet is Alpha Centauri, the closest star system to our solar system and which is only a bit over 25 trillion miles (40 trillion km) away.

Hot Jupiters

A lot of the planets that we were finding initially had been so-called hot Jupiters. These are gas giants, some far larger than Jupiter, that are orbiting even closer to their star than Mercury is to the Sun, maybe having an orbital period, or year, of something like 2 to 10 days. The reason that their name includes the word "hot" is because they can reach around 1,500 to 2,000 degrees Fahrenheit (815 to 1,100°C) on the side facing the star, with possibly 6,000 mph (9,700 km/hr.) winds distributing that heat around the planet. The relatively nearby star Vega (in the constellation Lyra) appears to host a giant planet so close to it that its orbit only takes 2.5 Earth days, and it is one of the hottest known, with a surface temperature of 5,400°F (3,000°C)[ccxlix]. Some of these hot-Jupiters seem to have a tail, like a comet, as gas and dust is being blown off of them, possibly around 10,000 tons (9,000 Megagrams) of material being blown off every second.

One system has a Neptune-sized planet orbiting a white dwarf (WDJ0914+1914) that is only a quarter its size. The star is evaporating the planet at a rate of 314 million tons (285 million Megagrams) per day[ccl]. In at least one case, a smallish, but extremely dense, rocky planet has been found close to its star, and that is believed to be the core of a hot-Jupiter after all of its gas had been blown off.

HD 80606 b is a highly unusual hot-Jupiter that is about 190 light-years away. It has a mass equivalent to about four Jupiters, but it also has a highly elliptical orbit taking it around its star (which is one member of a binary system) in about 111 days. As it gets closest to its star, it can heat up from around 1,000°F to 2,200°F (540°C to 1,200°C) in around 6 hours, producing storms with winds reaching speeds up to fifteen times the speed of sound. Then it flies far out in the system, cooling off rapidly. HD 37605 b is a similar gas giant that is three times Jupiter's mass with a highly elliptical 54-day orbit that takes it very close to its star.

The hot Jupiter world known as WASP-62b has no observable clouds or haze in its atmosphere (based on its spectrum) and has become the second known exoplanet expected to have completely clear skies[ccli].

Super-Earths

Another common type of planet that we have been finding is called super-Earths, which we mentioned when discussing Planet Nine. Not that such planets necessarily have an atmosphere, oceans, and continents like Earth, but they are rocky worlds like the inner planets of our Solar System, except that they are much larger, maybe ten times as big as Earth. This type of planet is so common that it has been speculated that our Solar System may have started off with that type of world as its original inner planets but Jupiter's trip through the Solar System might have driven some into the Sun, or thrown them out, and the inner planets we now have formed from the left-over matter from that chaotic time. Some of the super-Earth exoplanets are believed to be water worlds, covered by a deep ocean.

Being super-sized, super-Earths should be able to retain their heat better, sustaining their molten cores. That means that they will also retain their protective magnetic field for an extended period of time, so if life starts there it will have a long time to evolve and grow[cclii]. However, as an example of how un-Earth-like some of these super-Earths can be, Gliese 486 b is a rocky super-Earth 26 light-years away, 2.8

times as massive and 30% larger than Earth, orbiting an M dwarf star (smaller and cooler than the Sun) at 1.6 million miles (2.6 million km). Its year is 1.5 Earth-days and, being so close, it is receiving intense radiation from its star. It is also tidally locked. The average surface temperature is expected to be around 800°F (430°C), with a surface maybe similar to Venus, but the atmosphere is probably thin because the star's radiation will have blown most of it away[ccliii].

The super-Earth known as GJ 1132 b, 40 light-years away from us in the southern sky, is believed to have started out as a gas giant planet orbiting around a star in the constellation Vela. However, radiation from its sun burned off its outer layers of hydrogen and helium, leaving its rocky core as an airless super-Earth with 1.6 times Earth's mass. Now it seems to have gained a hydrogen-rich but oxygen-poor atmosphere thanks to volcanic activity[ccliv], although questions about its atmosphere still remain.

Red Dwarfs and their Planets

Red dwarf stars have been found to commonly have planets, and the nearest star to us, after the Sun, is the red dwarf called Proxima Centauri. It has a planet, Proxima Centauri b, that is in the star's habitable zone and may have oceans, but it is tidally locked to its star (like the Moon is to Earth), with one side always facing the star so that that side is probably a searingly hot desert, while the other side would be dark and frozen. There could be a twilight zone forming a band around the planet, between the hot and cold sides, where life might exist, with water pouring off glaciers that are finding their way from the frozen dark side. Being so close to the star, it is hard to assess much solid information about the planet at present, so whether or not it actually has an atmosphere is still unclear. What is known as the "Goldilocks zone" (where liquid water and maybe life, somewhat as we know it, might exist) is fairly close to a red dwarf because the star is relatively cool. Proxima Centauri seems to have at least two other planet[cclv]. Proxima Centauri b has a mass of about 1.27 Earths, Proxima Centauri c has about 7 Earth masses, while Proxima Centauri d has only about a quarter of Earth's mass[cclvi].

Red dwarfs are the longest-lived stars in the universe, so there would be plenty of time for life to evolve there. Red dwarfs also are very active stars, with dramatic flares blasting out from them regularly, so there could be some magnificent auroras illuminating the sky. These solar flares may put on a great display, but they can also be deadly to lifeforms, potentially stripping away a planet's atmosphere and irradiating the surface. There are indications, however, that the deadliest of the flares erupt from the poles of red dwarfs, while the planets orbit around the star's equatorial plane, so the planets shouldn't be in as much danger as we originally thought.

Far from being a hot-Jupiter, the Neptune-like exoplanet OGLE-2005-BLG-390L b has a surface temperature believed to be around -364°F (-220°C). With its core of rock and ice and a mass about five times that of Earth, the planet orbits at a distance of 2.6 AU (Astronomical Unit – the distance from the Sun to the Earth) around what is believed to be a cool red dwarf star that lies fairly close to the center of the galaxy. This was one of the most distant exoplanets that we had observed when it was discovered in 2005.

2MASS J23062928-0502285 is a red dwarf star, maybe twice as old as our Sun, but only slightly bigger than Jupiter, and it is far better known as TRAPPIST-1 (named after the Transiting Planets and Planetesimals Small Telescope). Its biggest claim to fame is its seven planets, 3 or 4 of which might be within the star's Goldilocks Zone, where liquid water could possibly exist on the planet's surface. All of the planets, which range from around the size of Mars to a bit larger than Earth, orbit closer to their star than Mercury is to the Sun, and all of them are expected to be tidally locked with one face of the planet constantly facing the star. The planets exhibit orbital resonance with one another which leads to regular interactions that may disrupt the tidal locking to some extent and also generate tidal heating that probably means some or all of the planets are geologically active. It is thought that the planets probably formed further out and migrated inward to their current positions. It has not yet been possible to confirm whether any of the planets has retained an atmosphere, but it is thought that the coronal ejections and strong solar wind make the retention of any atmosphere unlikely. However, the system is one that astronomers will be keeping a close eye on, especially since it is only about 40 light-years away[cclvii].

Other Exoplanets

There are a lot of other interesting exoplanets that we have found. There is a planet that we alluded to earlier that is largely solid carbon, like a diamond, possibly shining in space like a 30,000 mile (48,000 km) wide disco ball. It would be extremely hot and raining down silicate (glass) that is being blown by 4,000 mph (6,400 km/hr.) winds. Talk about being sandblasted!

370 light-years away is a young star, poetically named PDS 70, that is still surrounded by a cloud of debris known as a circumstellar disk, from which at least two gas giant planets have already formed. Other planets, including rocky terrestrial type planets, may also be forming within that disk of gas, dust, and larger debris. One of the gas giants (PDS 70 c) has been observed to have pulled some of the dust and debris around itself to form a circumplanetary disk. Some of the material from that disk will be pulled down to the planet to help it grow, but undoubtedly there are also moons forming within the disk. The planet is a few times more massive than Jupiter, so it could be forming a lot of moons. The other gas giant (PDS 70 b) also probably has a circumplanetary disk, but that hasn't been observed yet[cclviii].

Dust and debris is something else that can result from collisions, and that is what is thought might be the cause of the strange dimming recorded in relation to a binary star system named TIC 400799224 that is about 2,300 light-years away. Once every 19.8 days, something causes dimming in the system, but the amount of dimming can vary substantially, and sometimes it doesn't seem to occur at all. One idea for what is causing this is that two or more large bodies are colliding against each other on this regular cycle and creating a cloud of debris that quickly dissipates[cclix].

Globular clusters are compact, spherical collections of stars usually found in the halo of galaxies, and 150 of them are known to exist in the Milky Way. The closest one is M4, over 7,000 light-years away, sixteen hundred times further away than our nearest neighboring star and more than five times farther away than the majority of the exoplanets discovered. But we have found one planet, PSR B1620-26 (AB) b, in that globular cluster, orbiting 2 stars (thus the AB in the name). These stars are pulsars and the planet's gravity interferes with the timing of the radio pulses, allowing the planet to be identified. That is the same method that identified the first discovered exoplanets. Computer simulations suggest that, due to stars being packed close together in a globular cluster, any planetary systems there will also need to be compact. Planets that were further out would get ejected by the gravitational pull of passing stars, so, say, there would be no planets beyond something equivalent to Mars's orbit[cclx].

Earth-sized K2-141 b has an atmosphere of rock vapor, according to simulations. At 5,400°F (3,000°C) at its surface, its lava oceans evaporate, condense, and it rains down rocks. So, it's an Earth-sized planet but with a rock cycle instead of a water cycle.

Hot-Jupiters orbit very close to their stars, but some planets orbit far out. The brown dwarf Oph 98A has a planet-sized companion orbiting five times further out than Pluto is from the Sun. Their low masses and wide separation make them the weakest gravitationally bound binary system known. HD106906 b is circling two stars that are 336 light-years away from us and it has an orbit that is over 730 AU from its host stars, taking possibly 15,000 years for one circuit, and the orbit is tilted 30° from the orbital plane of the dusty disk surrounding its host stars[cclxi]. Presumably the planet formed closer in, but its orbit decayed until the whirling twin stars kicked it out to its present tenuous orbit.

80 light-years away, the white dwarf WD 1856+534 (only 40% the diameter of Earth) has a planet, WD 1856 b, that is seven times as big as its parent star and is orbiting it every 34 hours. Normally, planets and other bodies are expected to be destroyed or ejected from the system during the red giant phase and as the star ejects most of its matter[cclxii] before collapsing to a white dwarf, so the survival of this planet is intriguing.

Jupiter-type planets have been seen to have survived a main sequence star becoming a red giant and then collapsing into a white dwarf, but normally these planets are orbiting close to the white dwarf. Presumably their orbits have been disrupted by the changing star and the clouds of gas that would have been blown off. However, one such Jupiter-type planet named MOA-2010-BLG-477L b has been seen in what looks as if it could have been its original orbit[cclxiii].

Finding a planet around a white dwarf was kind-of surprising but finding one in a white dwarf's habitable zone was a bit of a shocker. Well, we haven't really found the planet itself yet, but what has been observed are dips in the star's brightness that are probably caused by a ring of about 65 comet- or moon-sized objects that are fairly evenly spaced out in their orbits. That spacing seems to imply that there is a planet a bit smaller than Mars orbiting slightly further out at a distance that would result in an ambient temperature on the planet of around 120°F (50°C). The star in question is WD 1054-226, which is around 118 light-years away[cclxiv].

Rocky world LHS 3844 b is tidally locked to its star, and one face is about 1,410°F (770°C) hotter than the other[cclxv], so it may have one face with erupting volcanoes while the other face is frozen solid.

Kepler-16 b was the first planet found to be orbiting two stars, although we now know that around half of the planets discovered are orbiting such binary systems. The planet is a gas giant that orbits well outside the habitable zone of either star.

Kepler-36 has two planets (b and c) that both orbit very close to their star and very close to each other. Kepler-36 b is a super-Earth with about 3.8 Earth masses while Kepler-36 c is a mini-Neptune gas giant with almost double the mass of its companion, and the two come within 1.2 million miles (2 million km) of each other at their closest point.

The Kepler-444 triple-star system has five planets ranging from around Mercury-size to that of Venus, and all of them are orbiting closer to their Sun-like star than Mercury is to the Sun.

K2-18 b is only 124 light-years away and became the first exoplanet known to have liquid water, an atmosphere, and a livable temperature around 99°F to 160°F (37°C to 70°C), similar to Phoenix, AZ (if you call that livable ;-) However, it appears to be a Neptune-like world, so no lakes and rivers, and no life-as-we-know-it.

Kepler-20 has two Earth-size planets in between three Neptune-size worlds. There is a possible sixth planet as well, but that's disputed.

Kepler-90 has eight planets ranging from Earth- to Jupiter-size, in the "right" size order around a Sun-like star (i.e., the gas giants orbit further out than the rocky worlds), but all are orbiting their star closer than Earth is to the Sun.

Mysterious exoplanets that appear to have densities far lower than our gas giants, about one-tenth the density of water, have been called "super-puffs". They could be gas giants that formed with a small and icy core that pulled the hydrogen and helium gas around it, and that gas expanded as the planet moved in closer to its star and heated up. Alternatively, they might be fooling us and are really exoplanets with expansive ring systems[cclxvi].

It was thought that a star system with more than about three solar masses total was unlikely to have a planetary system because the radiation from the stars would prevent planets from forming. However, a binary star system with a combined mass for the stars of about eight solar masses has been found to have a gas giant, about 10.9 times the mass of Jupiter, orbiting it. However, the planet, known as b Centauri (AB) b is orbiting the two stars at a distance of about 560 AU[cclxvii].

GW Orionis has three young stars in misaligned orbits around one another that have formed three distinct rings within the surrounding cloud of gas and dust around the system with the inner ring on an entirely different plane to the others[cclxviii]. As a consequence, any planets in the system could be substantially misaligned with one another. There is a large gap between the inner and middle rings, and it is suggested that the gap is caused by one or more planets orbiting the three stars at around 100 AU from the center of the system[cclxix].

The search for exoplanets has thrown up a number of mysteries, the best-known mystery being the one associated with the star KIC 8462852 that is better known as Tabby's Star (named after Tabatha Boyajian, lead author of the first paper about the star's strange light curve). Studying data from the Kepler Space Telescope, unusual dips in the star's brightness, up to 22% at one point, were noticed and they didn't seem to fit any regular pattern. A number of possible causes have been suggested, including a swarm of comets and even an artificially created Dyson swarm that aliens might be using to gather power from the star. As intriguing as some of the ideas are, especially that last one, further observations have ruled out some ideas and failed to provide support for others. The current idea that gains most support seems to be that the dimming is the result of the fragments of one or more shattered orphaned exomoons[cclxx].

Exoplanets are getting easier to find but are still rather hard to positively identify, and exomoons are even harder. Nevertheless, it looks as though at least one may have turned up in the Kepler data. Kepler-1708 b is a Jupiter-size exoplanet about 5,000 light-years away, and variations in the dips that it caused in the light from its star are best explained by the planet having at least one exomoon that is about 2.6 times the size of Earth. Kepler-1625 b is another exoplanet that is suspected of having an oversized exomoon, but not quite as big as Kepler-1708 b's[cclxxi].

Around 5,000 exoplanets have now been confirmed within the Milky Way, and now one may have been observed in the Whirlpool Galaxy (M51), which is 23 million light-years away. There is a binary star system there, M51-ULS-1, that consists of a neutron star or black hole that is sucking material off of its companion star which is about twenty times as massive as the Sun. That causes X-rays to be emitted, but for three hours the detectable X-rays dropped to zero. The best-fit model for that effect is that a gas giant about as big as Saturn passed between us and the binary system in an orbit about 19.2 AU out from the center of the system. It should come around again in its orbit in about 70 years, at which time we can check if that assumption is correct.

Rogue Planets

Elsewhere in our galaxy (and presumably in other galaxies as well) there are planets that have been thrown out of their solar systems and are now wandering the dark realms of interstellar space and are known as rogue planets. It is thought that rogue planets could outnumber stars in the Milky Way. They might be orphans lost to space when their host star became a red giant and was puffing away layers of their mass, or maybe they were ripped away by a passing star or flung out by a larger sibling planet or even by interaction with its star. Some rogue planets may actually be expired brown dwarfs that have used up their deuterium fuel. It is calculated that in the gas and debris around supermassive black holes, starless planets 3,000 times Earth's mass could form, that are called blanets, and which could have one-million-year orbits about 10 light-years out from the event horizon[cclxxii].

But although rogue planets are out in the extremely cold void of space, that doesn't mean that they must be cold. Earth is partially heated by its inner molten core, and most of the heat on Jupiter comes from the planet itself, not the Sun. From Jupiter, the Sun appears not much more than a star, although it looks a bit larger than the others. If there was liquid water and some energy source from inside such a rogue planet, life almost as we know it could exist there. Any lifeforms there would be unlikely to have eyes though, because there would be nothing to look at, since it would be so dark with no nearby star to provide illumination. If there are moons around exoplanets, they can also be heated by tidal effects, so we could have water worlds like Europa around rogue planets and other exoplanets.

About 90 or so light-years away is a pair of rogue planets, collectively known as the 2MASS J1119-1137 AB system. Each of the planets is believed to be about four times the size of Jupiter, and maybe larger. It is also believed that one of the planets has a moon about 1.7 times the mass of Earth, and that this moon would be getting about the same amount of radiation from the planet as Earth gets from the Sun, so there could be water and, dare we suggest, life on it[cclxxiii].

The search is on for orphaned exomoons, called ploonets, which are probably more numerous than exoplanets. They are imagined as having been cast loose from the chaotic dance between their parent planet (probably a hot Jupiter) and its host star. It is suggested that only about 10% of ploonets survive, the rest getting swallowed by the star, smashed into their parent planet, or get vaporized by stellar radiation.

Chapter 12: Life Begins

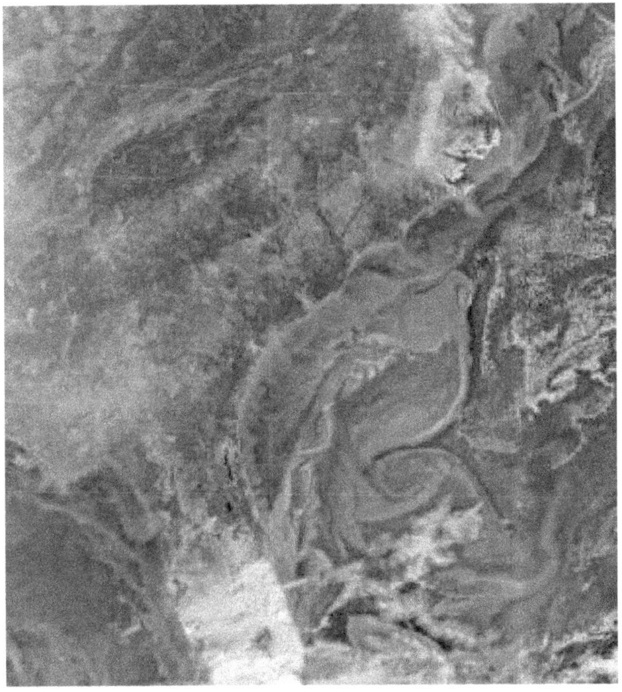

Organic Molecules and the Origin of Life

Supernovae create the basic ingredients of biology and spread them to worlds scattered throughout the universe. The atoms forged in the stars combined into molecules, sometimes helped by the radiation from the local star, and asteroids and comets have been found to have a lot of carbon-based molecules. Those are also known as organic molecules because they are the chemicals that are used by all of the living creatures that we know of. Methanol and other organic molecules have been spotted in gas clouds at the edge of the Milky Way, even though individual carbon and oxygen atoms are uncommon there[cclxxiv].

A study of five nearby star systems, that was looking at the chemistry in the protoplanetary disks, found from ten to a hundred times the amount of organic chemicals that was predicted by models. The organic chemicals were not evenly distributed throughout the disk, so it is suggested that where a planet forms in the disk could be important in regard to whether organic life emerges there or not. One of the most frequently detected organic chemicals was cyanide, which we tend to think of as a deadly poison that kills off life, but it is also believed to be a good chemical for getting life started[cclxxv].

Location is not only important for obtaining the organic ingredients, but also for providing the right conditions for life. In that regard, it appears that as many as one-in-ten star-systems have a planet of just the right size and distance from its star to sustain liquid water on its surface – a key requirement for life as we know it on Earth.

We, and other living matter on Earth, utilize lipids to forms the "sacks" (or cell walls) that hold our cells together, and we also use about 20 amino acids and a few nucleotides, which are all made up of those same organic molecules that appear all over the universe. Earth was bombarded with comets and asteroids containing such organic molecules during the period of the Late Heavy Bombardment, so organics could have arrived that way. DNA is a vital part of living cells, telling an organism how to put itself together and controlling its development over time, and all four of the building blocks (nucleobases) of DNA have now been discovered in meteorites[cclxxvi].

We could have also developed the hydrocarbons here. The atmosphere was probably mostly hydrogen, carbon dioxide, carbon monoxide and ammonia, plus steam and rock vaporized by the regular impacts back then. With the ultraviolet rays from the young Sun and no ozone layer yet to stop it, a hydrocarbon haze would probably have developed and then deposited the ingredients of life into the oceans. Volcanic activity may have provided some of the heat energy needed to let the organic chemistry take place. So, the building blocks of life were now on our planet from a number of different sources, and what was needed was some way of bringing them together and getting them working as a unit. There have been a number of ideas put forward as to how that came about. One of the first ideas was that the chemicals might have been pushed together in evaporating rock pools on the early Earth, but the high levels of ultraviolet radiation from the Sun at that time might have made that idea problematic.

Location, Location, Location

The favored idea now is that warm alkaline vents on the ocean floor probably provided the right conditions for life to start. There are vents that occur naturally from geological processes, but the impacts from asteroids pummeling the Earth during its early lifetime would have melted portions of the Earth's surface and heated nearby groundwater and could also have started such hydrothermal activity[cclxxvii]. These impact craters and their associated hydrothermal systems would have been more numerous than naturally occurring vents and would have provided some shelter for new microbial life from the bombardment. The temperature at those impact systems might have been too hot for life initially, but they would have cooled over time. These Hadean period impacts might have formed craters 1,000 miles or more in diameter. Such hydrothermal systems have been observed in the Chicxulub crater associated with the end of the dinosaur era, so earlier bombardments no doubt produced them as well.

The hydrothermal vents, from whatever source, would have an energy gradient from the warm water getting pushed up from below the surface, the water overhead would give protection from the ultraviolet light, and small cavities in the alkaline chimneys would form pockets where the chemicals could condense and become the first living cells. Experiments are being conducted to see how chemicals that are expected to have existed on the early Earth react when brought together in random combinations. Some evidence of molecular replication has been seen, and researchers are hoping to see chemicals react to produce their own catalyst, which would be considered a good step on the path to life[cclxxviii].

These first cells would have been different from basic

chemical mixes in that they developed metabolism. In other words, they found ways of taking in the chemicals they needed for energy and growth and had a means of getting rid of waste products. They also had to have some kind of template that guided the growth of the cell. We now have DNA doing that job, but almost certainly it would have been something simpler initially and might have been RNA. RNA is a very useful self-replicating molecule found in living matter and, although simpler, it is still quite complicated. Consequently, there was most likely something even simpler still to get life started, maybe something like PNA (peptide nucleic acid) which is an artificially created polymer similar to, but not quite as complicated as, RNA.

Once the living cells moved from the cavities in the vent funnels, they would have needed some form of cell wall to separate themselves from the surrounding environment. Lipids come together automatically in water and are what now forms the material of the cell walls, but the first cell wall membranes were probably mixtures of fatty acids and fatty alcohols. Experimenters have seen cell-like organic vesicles form automatically when such lipid precursors were added to a solution of organic material[cclxxix].

Chapter 13: Life Evolves

Single-Celled Life Branches Out

Life seems to have started on Earth as soon as our planet was cool enough to allow it, about 3.7 billion years ago[cclxxx], although there's some evidence that it may have started as early as 4.1 billion years ago[cclxxxi] (somewhere around September 18 in our "cosmic year"). That would have been during the period of the Late Heavy Bombardment, and the constant impacts from asteroids might have mixed everything up enough to allow the chemicals of life to arrive at the correct formula[cclxxxii]. So, somehow the Earth and the asteroids got life started, even if we have difficulty in working out exactly how the universe did it. In that very hot and active chemistry of the early Earth, it appears that a self-replicating molecule formed, leading on to what we call life. Then, for almost 2 billion years, life continued as single-cell creatures, but that life was learning to use different sources for obtaining energy and adapting to different environments.

Single-celled does not have to mean simple. We don't know how different the first single-celled creatures were from modern-day varieties, but some single-celled creatures we see today have a dozen or so cirri (or legs) that they can use to get around, yet they have no nervous system to coordinate their "legs". Instead, they have microtubules (which are also found in all complex cells) that connect one "leg" to another, and it is believed that these microtubules form a sort of mechanical computer to control the "legs" and the subsequent movement of the single-celled creature[cclxxxiii]. Evolution of lifeforms often seems to involve the accumulation of complexity, and that complexity doesn't arise from the number of genes we have but in the number of ways they can be expressed depending on which combination is activated[cclxxxiv].

Single-celled creatures can be very adaptable, which is aided by the available complexity. Scientists stripped a bacterium, that already had a smallish number of genes, down to what was considered the bare minimum of genes, each identified as being vital to its survival. It started out with a 901-gene genome and was left with only 493 genes. This minimal-genome organism survived and evolved to the point where, over 300 days, the organism had recovered about 80% of the fitness it had lost compared to the original bacterium[cclxxxv].

A slime mold is a big single-celled creature with no brain, yet scientists are studying it for clues to the evolution of intelligence because it can do some remarkable things. For instance, it can find its way through a maze[cclxxxvi], escape from a box, and sense and adapt to its environment. It also shows signs of memory, apparently learning from what it has done and where it has been.

We normally think of plants as being the first lifeforms to move out of the oceans onto the land, but bacteria like to explore too. There is growing evidence that 3.2 billion years ago, or more, microbes had been feeding off of the nutrients from volcanic flows that had found their way into estuaries or shorelines and then those microbes moved out of the water to feed on the cooling lava flows on land, enjoying the residual heat and the energy-rich chemicals[cclxxxvii]. If there is one special characteristic defining life, it might be adaptability.

First Life-Created Climate Crisis

In their constant adaptation to different energy sources, one type of creature (cyanobacteria) found that it could utilize sunlight for energy and carbon dioxide as food, and photosynthesis was born. This new green slime lifeform gave off oxygen as a waste product, which was really not appreciated by most of the lifeforms that were around back then. Up to that time, there had been very little oxygen in the atmosphere, but now it started to build up and it proved to be toxic to most lifeforms that were around at that time. This first great dying occurred around 2.4 billion years ago (October 29 in the "cosmic year"). To add to the problems, the reduction in carbon dioxide (which, as we are learning the hard way, is a greenhouse gas) meant that the Earth could not retain as much heat, and that resulted in the first snowball Earth, the whole planet being covered in ice and snow from pole to pole around 2.3 billion years ago (November 1 in the "cosmic year"). These early lifeforms reduced the carbon dioxide in the air and our activity is increasing it, but too much and too little carbon dioxide are both bad. We don't seem to have learned a lot since the time that we were all bacteria.

Life Spreads Worldwide

However, as we said, lifeforms are nothing if not adaptable, and some of them learnt how to use the oxygen as a very efficient energy component and gave off carbon dioxide in the process. Volcanoes also kept belching out carbon dioxide, so the level of greenhouse gasses rose again, even if not as high as before, and summers returned to the Earth. The ice retreated back towards the poles and some of the single-celled creatures learned that they could work together as sort-of "colony creatures", the way slime molds do.

The early single-celled creatures would have been what are known as prokaryotes, meaning that they were single-celled with no obvious nucleus or other organelles. It has been found that DNA in some modern-day prokaryotes can incorporate genetic material within themselves from other organisms, using a form of CRISPR gene-editing. So, they may be simple as lifeforms go, but they can do sophisticated things[cclxxxviii].

Around two billion years ago (November 9 in the "cosmic year"), eukaryotic cells (cells with internal "organelles") developed after one single-celled prokaryote-type creature engulfed another, and the two learned to live and collaborate together as one. The nucleus in a living cell is a very important organelle, providing overall control for the cell by storing its DNA, and mitochondria are another type which generates a source of chemical energy for a cell. This mixing and merging led to a lot of variety, pushing evolution faster.

Around 1.5 billion years ago (our cosmic calendar now reads November 22), eukaryotes divided into three groups, one leading to plants, one to fungi, and one to animals, although all three groups were probably single-celled at that time, even if some of them might have been collaborating in groups. Animal life, in the form of sponges and/or soft-bodied comb jellies, may have existed by around 890 million years ago (December 8 in the "cosmic year"), according to some fossil evidence[cclxxxix]. It has been suggested that around 800 million years ago, or maybe earlier, a small genetic change in the GK-PID molecule enabled real multicellular creatures to develop, rather than just collaborating groups of single celled creatures.

Around 720 million to 635 million years ago (December

12 to 15 on the cosmic calendar), the Earth went through another ice age or snowball Earth event. Then around 630 million years ago, animals developed bilateral symmetry, showing a definite top and bottom and front and back, and the first creatures like this were probably some kind of worm. The fossil record indicates that around 600 million years ago more complex animal life was evolving, and around 570 million years ago (December 16 on the "cosmic year" scale) we see vertebrates (animals with backbones) start to develop from invertebrates (those without a backbone).

Oxygen levels were still fairly low, which wasn't too much of a problem for ancient animal groups like sponges. Around 550 million years ago (December 17 in the "cosmic year") we see a rise in oxygen levels that really enabled more complex animals to develop, and we see life taking off in the oceans, and plants began to move onto the land.

We normally think of oxygen being produced via photosynthesis, but it seems that life has come up with at least one alternative method. Microbes have been found in the depths of the ocean where sunlight doesn't reach, so photosynthesis cannot take place, yet they are generating oxygen somehow. Exactly what the process involves is unclear at this point, but life finds a way[ccxc].

Microfossils have been found of a creature (Bicellum brasieri) that shows developments towards living beings with a skin that is distinct from the cells inside the body[ccxci]. The creature was only a few tens of micrometers across and consisted of a few dozen cells. It had a central ball of tightly packed oval cells surrounded by an outer layer of sausage-shaped cells, and some of the fossils only showed the central ball, which may indicate a larval stage. It is believed that the oval cells divided to create the ball and then formed the elongated cells that migrated to the surface to form the creature's "skin". Single celled creatures found that they needed cell walls to separate themselves from the surrounding environment, and multicellular creatures needed skin, scales, or an exoskeleton for the same purpose.

Around 535 million years ago, fish and other aquatic life really took off in what is known as the Cambrian explosion. By 500 million years ago (December 18 in our "cosmic year"), animals had started venturing out of the oceans to explore the land, probably as a result of tidal pools intermittently drying out. Around 489 million years ago, we had the Great Ordovician Biodiversity Event, and by 465 million years ago, plants started taking over the land in a big way. The oldest known insects are seen around 400 million years ago (December 21 on our cosmic calendar), and some plants started to develop woody stems, probably as a form of defense against the insects.

The first tetrapods (4-legged creatures) developed around 397 million years ago, probably in shallow freshwater. Around 350 million years ago, oxygen levels were about as high as they were going to get, allowing massive dragonflies and the like to exist. By 340 million years ago (late on December 22 in our cosmic scale), amphibians were branching off from other tetrapods. Some tetrapods moved back to an aquatic life again, showing that evolution can reverse course sometimes[ccxcii]. 310 million years ago, the remaining tetrapods split, with sauropsids leading on to reptiles and dinosaurs, and synapsids leading eventually to mammals, passing through "dinosaur-like" creatures such as Dimetrodon along the way.

By 225 million years ago (late on Christmas Day in our "cosmic year"), dinosaurs began developing feathers to provide themselves with heat insulation. This seems to have been mostly the smaller dinosaurs and maybe the young of some of the larger ones[ccxciii]. By 130 million years ago we had the first flowering plants, and by 100 million years ago (December 29 in our "cosmic year"), dinosaurs reached their peak in the Cretaceous period, so they may not have been able to smell the roses, but they could smell some flowers.

85 million years ago, modern primates split from other mammals, leading on to humans starting to diverge from chimpanzees and bonobos about 6 million years ago (December 31 at 8:10 p.m. using our cosmic year scale) as bipedalism emerged for the first time.

Humans Take Over

Australopithecus are dated from about 4 to 2 million years ago (9:26 through 10:42 p.m. on December 31, per the "cosmic year") and seem to have been caring for their sick and injured. There are a number of varieties of Australopithecus. Footprints at a site in Laetoli, Tanzania, from 3.66 million years ago are thought to have been caused by Australopithecus afarensis, which is the early ancestor of humans that included Lucy. Lucy died about age

3, and could be described as part ape, part human, and she was built for climbing trees. Another set of footprints from the same region and time seems to have been made by a different variant of Australopithecus[ccxciv].

Australopithecus sediba, based on an analysis of its spinal bones, spent a lot of its time walking on two legs. They lived about 2 million years ago, and their remains were found in the Malapa cave system in South Africa. This type of hominid was also adapted for tree climbing, making it appear to be halfway between the tree-dwelling apes and the bipedal humans[ccxcv]. For anyone who has suffered from back pain, the move to walking on two legs may seem as if it was a bad idea, but it did let us see further and spot predators from a safer distance and it freed up our hands so that we could make better use of tools.

Homo habilis evolved around 2.3 million years ago, surviving through to about 1.4 million years ago in eastern and southern Africa, and seems to have used stone tools. Its brain was about the same size as that of a chimpanzee but seems to have had more development in the frontal cortex region[ccxcvi].

Homo erectus, which lived in Africa, western and southern Asia, and maybe Europe, from around 1.9 million to 110,000 years ago (i.e., they died out about 5 minutes to midnight on December 31 of the "cosmic year"), and they were the first to develop shorter arms and longer legs and to move outside of Africa. They developed a powerful grip that could be used with precision, which was good for making more complex tools, and its use also shows up in the etched patterns on shells that indicate that they had a form of artwork. With a brain that was about halfway between a chimp and a modern human, there are significant indications that it probably used language and fire. Language and toolmaking could be related as far as the brain is concerned, with both using similar areas in the left hemisphere of the brain[ccxcvii].

The use of fire was very important in the development of humans, and it's a "tool" that no other animal seems to have discovered. Cooking food made it easier to digest and provided more energy, which allowed our energy-hungry brains to get bigger. The campfire also became a focal point for communities to gather around at night, and later people started to discover that fire could transform things other than food, and tools were suddenly no longer limited to stone objects but could be made from metals[ccxcviii]. Now humans were not just surviving off of the natural world but were starting to transform it, and I'll leave you to decide if that was a good development or not.

Homo heidelbergenis lived from 700,000 to 200,000 years ago, and they were found in eastern and southern Africa, Europe and maybe Asia. They made clothing, used fire and built shelters (as did Neanderthals) and that enabled them to move into colder regions. They were the first human species known to live in cooler climates and regularly hunt large animals. It is likely that they evolved into Neanderthals in Europe and into Homo sapiens in Africa.

Homo neanderthalensis (Neanderthals) lived from 400,000 to 28,000 years ago (11:43 to 11:58 p.m. on December 31 of our cosmic calendar) in Europe and southwestern to central Asia. They were shorter and stockier than us, with protruding brow ridges and brains as large or larger than ours and they were better adapted for cold conditions than Homo sapiens. They drew on cave walls, showing that they used forms of art, and they apparently cared for their sick and injured.

Denisovans lived somewhere around 300,000+ years ago to 15,000 years ago, but there are few fossil records, so the dates are uncertain. They may have evolved from members of Homo erectus and are known to have lived in east Asia, and they overlapped and interbred with Neanderthals and Homo sapiens. A skull that is at least 160,000 years old has been discovered at Xujiayao in northern China, and it is thought that it could be Denisovan. If so, it would indicate that they had a brain that is a bit larger than Homo sapiens[ccxcix].

Homo naledi were roaming regions of southern Africa around 335,000 to 235,000 years ago. They had a mix of ancient and modern features, such as a small brain and very long fingers, but modern human-like hands and feet. Seemingly, they could walk on two legs but also climb and hang from trees. It appears that they were already carrying out something akin to funeral rights for the dead 250 thousand years ago, if the discovery of the skull of a child of the species is being interpreted correctly[ccc].

Homo luzonensis are known to have existed up to 50,000 years ago, but when they appeared on the scene is obscure. Their remains were discovered on the island of Luzon in the Philippines, and they were diminutive humans with a mix of ancient and modern features. It's thought they might have evolved from an early pre-Homo erectus dispersal out of Africa. They thrived in woodland environments.

Homo floresiensis (also known as the hobbit people, about three feet tall) lived from around 190,000 to 50,000 years ago on Flores Island, Indonesia. They had tiny brains, large teeth, no chin, receding forehead, and they are known to have used stone tools, hunted small elephants, and they may have used fire. It's possible that they shared a common ancestry with Homo luzonensis and, like them, they thrived in woodland environments.

Homo sapiens made their appearance around 300,000 years ago (13 minutes before midnight at the close of our cosmic year) and we have not killed ourselves off yet. We had taken on a form that would be virtually indistinguishable from modern-day humans by about 130,000 years ago (6 minutes before midnight on December 31 of our "cosmic year"). We moved out of Africa around 100,000 years ago, became the dominant human group by around 30,000 years ago, and by 15,000 years ago were the only humans left and had spread around the world by about 13,000 years ago[ccci] (less than one minute before midnight at the end of our "cosmic year"). Our seeming vulnerabilities, such as being dependent on others and feeling compassion and empathy, might have helped. We weren't the biggest and strongest, so we had to learn to cooperate and work together. Neanderthals and Denisovans also made tools, but we lived in larger groups than others and formed alliances beyond our immediate group. We learned to generalize with specific populations specializing in certain environments, and our emotional neediness drove us to connect to others. Care led to longer lives and more ability for the elderly to pass on wisdom to their ancestors. Cooperation allowed them/us to better survive climate changes and learn new skills from one another. It is thought that genetic changes may have led to more trusting behavior[cccii]. What gave us our insatiable need to feel special, and superior to every other creature, is anyone's guess.

Anyway, we are the sole survivors of seven or more types of humans we shared the planet with during our time on Earth. It seems that we probably didn't kill off the others, but it was attributes like compassion, tolerance and the desire to make connections that helped us to survive the changes that the world threw at us. That desire for connection is also probably what led to us interbreeding with a lot of the other types of humans. Perhaps it is appropriate that the Flintstones and the hippies with their "make love not war" motto both rose to prominence at the same time in the 1960s.

But it does seem that our cooperation motives were driven by our own need for survival. Although the elderly might have been given assistance, senicide (the killing off of people who were past their productive lifespan) was fairly widespread in the past, and as soon as people were reasonably secure about meeting basic needs, they appear to have started squabbling over the little extras that indicated status. The selfish gene may not be a complete myth after all.

Mass Extinctions & Adaptability

The Earth went through a number of traumatic events, including the asteroid strike that wiped out the dinosaurs (except for those that we now call birds). But life always proved adaptable, and the removal of one type just meant that a different type had an opportunity to grow. There is a saying about nature abhorring a vacuum, and life seems to show the need to fill any available niche.

2,400 million years ago, there was the Great Oxygenation Event that we mentioned earlier, that was caused by rising oxygen levels due to photosynthesis[ccciii]. That was very bad for most of the early lifeforms, but it gave the planet and the later lifeforms a much better source of energy, without which it is hard to see how advanced forms of life, such as reptiles and mammals, could have evolved.

542 million years ago, there was the End-Edicaran extinction, which was the result of an anoxic event (where the ocean got depleted of dissolved oxygen) and that event is related to the rise of animals (the first multicellular organisms) which led to another altering of the environment.

The Ordovician radiation or Great Ordovician Biodiversification Event (GOBE) occurred around 467.5 million years ago and saw the Cambrian animal-types largely replaced by the Paleozoic animal-types. Both tectonic activity and global cooling have been suspected of playing a part in the killing off of the old types, allowing the new to evolve to fill the vacated ecological niches, but there could have been a cosmic connection too. Meteor hunters have identified that there was a heavy influx of space rocks around that time, maybe a hundred times as intense as usual. This could easily have led to local extinctions and have triggered GOBE. The incoming space dust might also have caused, or exacerbated, the ice age that occurred around that time. The space dust and meteors are believed to have come from the so-called L-chondrite parent body event. That was when a huge asteroid in the asteroid belt is believed to have shattered in a major collision with another body, and many of the meteors that we record today are also believed to have come from that event.

445 million years ago there was the second worst extinction event, and that is known as the Ordovician-Silurian extinction. That may have been caused by a gamma ray burst or global warming related to volcanism, along with anoxia. Or perhaps a combination of them all. A gamma ray burst might have directly killed off life on land and near the surface of the oceans while, at the same time, destroying the ozone layer allowing the Sun's ultraviolet rays to do more damage but also create more mutations speeding up evolution[ccciv].

372 million years ago was the Late Devonian extinction event that seems to have been related to volcanism, and possible also to one or more impactors. This seems to have been a series of extinctions, and also included widespread oceanic anoxia.

The worst extinction event occurred about 251.9 million years ago and is referred to as the Permian-Triassic extinction. This event wiped out a lot of species including trilobites who had made their appearance around 542 million years ago and were the first to develop image-forming eyes. But death is part of life and as a consequence of the extinctions, sauropsids (particularly dinosaurs) became dominant, and synapsids, the ancestors of mammals, became small, nocturnal animals. This extinction event seems to have involved volcanism, an impactor, an ice age, and an anoxic event related to the slowing of ocean circulation and rise in greenhouse gasses. It may have begun with dramatic volcanic eruptions in Siberia that blocked the Sun, cooling the climate and disrupting ecosystems, while later eruptions added greenhouse gasses and created global warming, causing more disruption to lifeforms[cccv].

201.3 million years ago, another mass extinction at the end of the Triassic let dinosaurs become even more dominant, and proto-mammals became warm-blooded (maintaining their internal temperature). Again, this event seems to have involved volcanism and an impactor.

The extinction event that we all know about occurred about 65 million years ago and is called the Cretaceous-Tertiary extinction or the K-T event. This is the one that wiped out the dinosaurs when the 6-mile-wide Chicxulub impactor hit, forming a crater hundreds of miles across and blasting millions of tons of red hot or molten rock into space. As the expelled material fell back to Earth, it resulted in widespread destruction with forests burning worldwide and led to something resembling a nuclear winter, and then there was volcanism on top of it all. But the event also made it safe for hominids (of which humans are a part) to develop.

Fossil finds from a site in North Dakota seem to imply that the Chicxulub impactor hit during late spring or early summer in the northern hemisphere. However, there are questions about whether this event was totally responsible for the demise of the dinosaurs, because there are indications that dinosaur numbers had been declining for about 10 million years before that, and the impactor may have just been delivering the final coup-de-grace[cccvi].

Currently we are living through the Holocene extinction that started about 11,650 years ago, and which is mostly due to human actions leading to destruction of habitats and climate disruption resulting from the burning of fossil fuels and other activities increasing greenhouse gasses. Some prefer to see the Holocene as having been replaced by the Anthropocene, starting around the 1950s. That was when nuclear weapons testing started affecting the globe and industrialization, with its coal and other fossil fuels, really took off in a big way, leading to our current crisis[cccvii]. It is estimated that the current extinction rate is 100 to 1,000 times higher than the natural background extinction rate.

Mass Extinction Cycles and the Cosmos

There have been other smaller mass extinctions as well and it was noted that mass-extinction events followed an approximately 26-million-year cycle, and a significant periodicity of around 30-million-years showed up in an analysis of documented impact craters. For example, apart from the Chicxulub crater, we have the 56-mile (90 km) diameter Popigai crater in northern Siberia that dates from about 36 million years ago, close to the extinction event at the end of the Late Eocene[cccviii]. The impactors for these events may have come from Earth-crossing asteroids or icy comets from the Oort Cloud, but normal asteroid hits seem unlikely to follow any particular timing pattern. Also, a passing star could disrupt the Oort Cloud, but again we'd have to ask why the periodicity? What has been discovered is that the solar system oscillates with respect to the midplane of the Milky Way on a period of about 60 million years, so it passes through that plane every 30 million years approximately. With possible dating errors, etc., the cycles seem to coincide well. A large fraction of our galaxy's normal matter resides in the flattened disk and a close encounter with a large cloud of gas and dust or dark matter could send a comet shower heading in our direction.

We have just passed through the midplane of the galaxy from "below" and are still close to it, and it takes more than a million years for a comet to fall from the Oort Cloud to the inner solar system. Cycles of tectonic activity and mountain building that drive sea level fluctuations have also been suggested as following a roughly 30-million-year cycle, although most geologists think these events are random. Geomagnetic reversals are suggested by some to show an approximately 30-million-year cycle. These cycles have been much debated and whether there is really any significance to them, or if they are even real, is still to be fully determined.

We do know of more regular-style objects that will have disrupted the Oort Cloud, although not on a periodic cycle. 70,000 years ago, Scholz's Star (a binary star system, one of them being a brown dwarf, the other being a red dwarf with about 15% of the Sun's mass) passed through the

Oort Cloud and in 1.3 million years' time, Gliese 710 is likely to pass by us at a distance of 17,000 AU, again well within the bounds of the Oort Cloud. Gliese 710 has 66% the mass of the Sun and will likely dislodge a considerable number of comets, with some probably heading our way[cccix].

Another suggestion for the periodicity noted is that Earth could capture dark matter particles that then accumulate in the planet's core, growing in number until they undergo mutual annihilation, producing prodigious amounts of heat in Earth's interior, maybe producing five hundred times Earth's normal heat flow, and fueling the regular pulses of geologic activity. Such geologic events are linked to mass extinctions, but we still don't know what dark matter is and what it is capable of beyond gravitational attraction, so this kind of idea falls firmly in the realm of speculation.

Sometimes it seems that the cosmos is out to kill us in many different ways. 2.6 million years ago there was the Pliocene marine megafauna extinction event that seems to have been caused by a supernova going off only about 160 light-years away[cccx]. The cosmic rays from the supernova hit our atmosphere sending a cascade of muons (like very heavy electrons) raining down on the planet. Although most muons pass through living tissue with little or no effect, this influx would have been a dangerous dose, and large sea creatures living in shallower waters would have been seriously affected and it is these creatures that went extinct. Had the supernova occurred closer to us, it might have ripped away our atmosphere and led to total global extinction.

Chapter 14: Search for Life in the Universe

Follow the Water and the Oxygen

Something like two trillion galaxies are estimated to exist in the visible universe, with about 100 billion stars (on average) in a galaxy (many of the galaxies are dwarfs), so that gives an awful lot of stars. We are also finding that it is very common for stars to have at least one planet, and often many more. It seems highly unlikely that, with all the available territory, that we're the only system with life.

"Follow the water" has been the mantra in the search for alien life, because liquid water seems to be essential for life-as-we-know-it. However, just looking at our solar system, we see that not having liquid water on the surface doesn't mean there's not liquid water there. Planet Earth is the only one in our solar system with readily available water on the surface and, true, it's the only planet we know of that has life, but Mars is believed to have water below the surface, and we also have a considerable collection of ice worlds that are known or suspected of having vast liquid water oceans below their icy surface. There would be no way to know that such places had available liquid water from the kind of distances we are dealing with in regard to exoplanets.

"Follow the oxygen" is another path that has been suggested because our atmosphere didn't contain noticeable amounts of oxygen until cyanobacteria made their appearance about two and a half billion years ago. Prior to that, the magma and other geological activity had kept oxygen locked up fairly well. However, there are other ways that good amounts of oxygen could end up in the atmosphere of a planet without life playing a part. If the surface of an Earth-sized planet was covered with around fifty times the volume of water that we have, it would exert enough pressure on the crust to dampen down (pun intended) geological activity, and so the oxygen would not get locked up but instead would build up in the atmosphere. Secondly, a planet that is a hot desert world with little water and which has a solid surface would start out with a lot of water vapor in the atmosphere that would get broken down by the UV radiation from its star, releasing oxygen. Once again, with the solid surface being unable to remove the oxygen, it stays in the atmosphere. The third possibility is where the planet starts out with excess carbon dioxide leading to a runaway greenhouse effect. That would again lead to water vapor in the atmosphere which breaks down to leave oxygen, and the surface of the planet would be so hot that it would prevent the chemical reactions from occurring that might otherwise tie up the oxygen. So, looking for water or oxygen may be good starting points in the search for life, but they don't give definitive results[cccxi].

We tend to look for life-as-we-know-it, but life on Earth is the product of a particular set of circumstances that it has adapted to. We have seen that life adapted to the changing atmosphere after cyanobacteria invented photosynthesis and, whatever our planet or the universe has thrown at it, life on Earth has found a way to cope with it and even thrive despite or because of it. There's no reason to believe that life couldn't have adapted to environments far different from Earth, especially when you see that life has adapted to and thrived in some of the most extreme conditions here on Earth.

People have speculated about life in the universe for 2,000 years or more, even before we had identified planets in our solar system as being worlds like ours (they were thought of as wandering stars before that). We see that life-as-we-know-it relies heavily on water, and we know that liquid water can exist in strange places such as many of the moons of the gas giants, and maybe even Pluto. So, perhaps Earth is the strange place – it's the only example in our solar system where liquid water exists in reasonable amounts on the surface, yet we can point to many places that have underground oceans that should be able to support life.

We can expect life to be made of elements that are fairly common at the specific location, and the ten most abundant chemicals are hydrogen, helium, oxygen, carbon, neon, iron, nitrogen, silicon, magnesium, and sulfur. Silicon is more common on Earth than carbon, but the latter is more useful chemically and it combines more readily with other chemicals. Consequently, carbon, hydrogen, nitrogen, oxygen, phosphorus, and sulfur (often abbreviated as CHNOPS) form the basis of life here and that is what we are mainly looking for elsewhere. The elements that life uses are light elements, so they should be more abundant in the far reaches of the solar system (where the moons with subsurface oceans exist) than they do in the inner region of the solar system where Earth is. As mentioned above, you will often hear the expression "Follow the water" in connection with the search for extraterrestrial life because liquids, especially water, carry nutrients into and out of cells.

On Titan, lifeforms on the planet's surface might have to use liquid methane and ethane instead[cccxii] because water is only available on the surface in the form of ice. It was expected that hydrogen, resulting from ultraviolet sunlight breaking apart acetylene and urethane in Titan's upper atmosphere, would be distributed fairly evenly throughout Titan's atmosphere, but it was found that hydrogen levels drop off near the surface. It was also expected that Titan's surface would be coated with acetylene, but the surface seems to be free of it. One suggestion is that a methane-based lifeform could be using the acetylene as a food source and using up the hydrogen the same way that we use up oxygen. Of course, there are other suggested non-biological processes that might account for these acetylene and hydrogen values as well.

Icy Moons and Dwarf Planets

It is believed that liquid water oceans exist beneath the icy outer shells of the moons Europa, Ganymede, and Callisto (around Jupiter), along with Mimas, Enceladus, and Titan (around Saturn). There is also Miranda and Titania (around Uranus,) and Triton (orbiting Neptune), as well as Pluto, and many other potential sites for subsurface oceans. That seems to make it possible for life to exist in all of those places, but there are some other issues that need to be accounted for too.

In order for the correct mix of chemicals to get dissolved in the water, you need interaction between the ocean and a rocky sea floor. With Pluto and giant-sized moons such as Ganymede (the largest moon in the solar system), Callisto and Titan, the pressure in their oceans may be so great that forms of ice can occur that sink in water, and that could end up covering the rock and preventing the necessary interaction. Still, it's fun to think of fish or squid-like creatures swimming around in all those subterranean oceans.

Life on Mars

The two 1976 Viking landers each carried out the "labelled release experiment" which involved adding Martian soil to a damp, nutrient-rich, radiation-tagged substance and later analyzing the resulting gasses. The aim was to see if any possible Martian microbes would metabolize the nutrients and give off gas containing the radioactive tracer, and that seemed to happen. In fact, the results came back more positive than expected. Trying it again with a sterilized sample of soil left no radioactive tracer changes in the air. People got excited about having found evidence of life, but another experiment testing for the presence of organic molecules came up blank. In the end, NASA said the tests were negative for life, but were the result of some strange, unknown chemistry[cccxiii]. However, "strange unknown chemistry" sounds like a reasonable description of life.

Since then, it has been shown that Mars definitely had conditions suitable for life-as-we-know-it in the past, and ancient lake beds have been explored and organic molecules found, but none of that proves that life is or was there. Some of the organics from Mars (specifically those found on a meteorite that came from Mars) have been determined to have been caused by water interacting with the volcanic Martian rock. However, such organics could still have been used to build lifeforms[cccxiv].

NASA's Curiosity rover recorded a background level of around 0.41 parts per billion of methane in the Martian atmosphere, but on at least six occasions it noticed spikes in that level, going up to around 4 parts per billion and in one case 10 parts per billion. The methane seemed to be coming from inside Gale crater where Curiosity is, and it is thought that the methane might have leaked out of some kind of crack in the surface, or just maybe it's resulting from a biological source. With methane having a detectable lifespan of only approximately 300 years, finding any at all there is a bit of a surprise[cccxv]. However, it has been discovered that exposing some meteorites to UV radiation can cause methane to be released, and there are plenty of meteorites littering the surface of Mars and very little to stop the UV light from the Sun from reaching them, so it doesn't need Martian lifeforms for the presence of methane on the planet.

If there is life on Mars, it will have to contend with deadly UV radiation because the planet lacks a magnetic field and a thick atmosphere of the kind that protects life on Earth. However, there are protected areas on Mars. Mars has caverns that are believed to have been lava tubes, and it is calculated that the UV radiation inside these tubes is only about 2% of that on the surface. That could be good for any indigenous lifeforms and also for any explorer from another world[cccxvi].

Life on Venus

Our sister planet, Venus, is believed to have had liquid water on its surface and to have been conducive to life-as-we-know-it in its early existence. It has been suggested that the surface of the planet might have been inhabitable for around 2 to 3 billion years. However, another study looked at the effect of the planet probably not having plate tectonics recycling surface minerals, and that could have reduced the habitable period to around 900 million years[cccxvii]. Another suggestion is that any H_2O on the planet was never more than steam or water vapor in its atmosphere, and no liquid water was ever on the surface.

It has been thought that life could exist in the temperate clouds although the concentration of water in Venus's clouds may be only one hundredth of that required by the most resilient microorganism on Earth, according to one study[cccxviii]. Now there are claims, disputed by some, that an inexplicable surplus of the rancid gas called phosphine, at 20 parts per billion, has been found in the clouds. On Earth, most phosphine is produced by microbes, but it can be created abiotically under great temperatures and pressures[cccxix], and Venus is not short of either of those. It has also been suggested that the phosphine claimed for Venus's atmosphere might be sulfur dioxide[cccxx].

Anyway, if phosphine has really been detected in the clouds of Venus, that leaves us with the options that it may be from some unknown chemical process or from microbial life[cccxxi]. Also, we know that Venus has lots of volcanoes, but we don't know how active they are. If they are as active as some volcanic regions on Earth, the observed phosphene gas could result from phosphorus being ejected from the planet's core in such volcanic eruptions[cccxxii].

Are We Alone?

The potential alien lifeforms we've discussed so far are nothing much more than bacteria or maybe fish or octopi. Some of those may even be called intelligent, but what we really want to find are technology-capable creatures that are out there trying to understand the universe, similar to us. Maybe something like the Klingons or Romulans of Star Trek fame. OK, that might make it seem lucky that we haven't come across any "little green men" yet. Our own history of exploration and discovery on Earth, and what has, way too often, been the fate of indigenous populations when more powerful "civilizations" discovered them, should be something of a warning to us.

There are, of course, incredible distances to cross to get from one solar system to another, but maybe some thought the Atlantic and Pacific oceans would be uncrossable too in the past. It is just possible that we are being visited by aliens who have their own version of Starfleet's Prime Directive and therefore conceal themselves from us as best they can. Some people might point to the record of UFO sightings as evidence of such activity, and apparently about a third of US citizens believe they have seen alien spacecraft, but people tend to forget that the U in UFO stands for "unidentified". Calling something a UFO doesn't identify it as an alien spacecraft, it says that we don't know what it is. I have seen two things that could be described as UFOs, one of which I have come up with two plausible and mundane possibilities for, while the second does remain unexplained to me, but I'm not going to jump to conclusions about it and could never prove anything about it one way or another.

The well-known Drake Equation tries to estimate how many intelligent civilizations we are likely to encounter in the Milky Way. It does that by first taking the number of stars in our galaxy, multiplying that by the percentage of those with planets and the average number of planets that such a star would have. That gives a count of the real estate locations where life might have started. Then the formula considers the percentage of such planets (and perhaps we should include moons now) that do develop life, and then what percentage of those will go on to evolve intelligent life (however you want to define "intelligent"). Finally, we have to consider what percentage of intelligent life will produce technology that we can get signals from (identifiable from the signals that the universe produces by itself), and then we have to take into account how long such signals would be observable by us. We would not be able to observe their signals if their civilization wiped itself out some way, or if they moved on to a technology that we couldn't detect[cccxxiii]. We are getting to the point where we can supply some reasonably accurate figures for some of the variables in the equation, but we still only have wild guesses for others.

I used to participate in the SETI-at-Home program, and the fact that we have been listening for transmissions from alien sources for over half a century and not heard anything definitive could have any number of reasons. It could be that intelligence is common throughout the universe, but that technology is a bit of an aberration. It could be that "intelligent" technological societies end up killing themselves off before they have much time to consider talking to the universe. Perhaps they've found better ways to communicate than by using radio waves. Maybe they're just not interested in us[cccxxiv]. Or perhaps they are not out there and we really are alone, but that wouldn't mean it's not worth listening to see if they are there or not.

Life As We Don't Know It

We have mostly considered where microbes and small-ish animals not totally dissimilar to those on Earth might be found. We've also looked briefly at lifeforms with different chemical makeup, as would be required if life exists on the surface of Titan. Such Titan lifeforms have been imagined as being enormous single-celled beings like large sheets floating near the edges of the methane lakes and breathing in hydrogen[cccxxv].

On Earth, lifeforms have developed and adapted to utilize just about every available environment, including those that at first glance would seem to be totally inhospitable to life. We find extremophiles enjoying toxic waste and living in regions of heat and cold that would be intolerable to other forms of life. There are even the tiny tardigrades that have survived in the near vacuum of space for a while.

Life on Earth has gone through a number of iterations following the various mass extinctions, but the new lifeforms have always been recognizable as animals or plants of some sort. Of course, all these lifeforms have grown from the same family tree, inheriting genes from lifeforms that went before and have been shaped by the same planet, even if conditions on the planet have changed substantially over time. But planets and moons come in a vast range of sizes and locations, leading to variations in temperatures, the strength of gravity, the density and transparency of the atmosphere, etc., etc., all of which the lifeforms will have to adapt to. Animals on Earth have either bones or exoskeletons to support them against gravity, but on a planet with a really strong magnetic field there will also need to be some kind of shell to protect the animal from that radiation, and ways to repair the almost inevitable radiation damage[cccxxvi].

When we get to planets like our gas giants, which seem to make up a lot of the planets throughout the universe, there have been ideas for lifeforms there too. The possible lifeforms are imagined as "swimming" through the gaseous body of the planet, some feeding off of the gases, maybe congregating as herds of gigantic jellyfish-like creatures with other creatures being predators who feed off of these herds.

Larger terrestrial-type planets, such as super-Earths, would have higher gravity that would result in shorter, stockier animals who might crawl along the ground. Planets not much bigger than Mars that have lower gravity than Earth and yet could retain a reasonably dense atmosphere, would make it much more convenient for creatures of nearly all sizes to develop flight, even if it is mostly soaring on thermals and gliding.

We see carnivorous plants on Earth and there seems to be nothing to prevent plants adopting other animal-like behaviors, such as moving around under their own volition if there were sufficient energy resources readily available and it met an evolutionary advantage.

Looking for life is looking for the unexpected, or what you wouldn't expect to find just from anticipated, or at least possible, chemical reactions. And maybe we shouldn't limit life to the surface and oceans of planets and moons. We know that tardigrades can survive for a while in space, so maybe life could have evolved there too.

A Living Universe Maybe

String theory comes up with 10^{500} possible universes, but most of those would have laws of physics that make them incompatible with life, and many even incompatible with stars or matter itself. The question has been raised as to how our universe came up with the set of laws it has, which make things work out well. It seems as if our universe was fine-tuned for life, but others suggest that all 10^{500} universes came into existence (which is one form of the multiverse) and we're here because this is the one that was able to give rise to us. That says that we are only here because we hit the jackpot, being one universe in a multiverse where not all are survivable for matter, let alone life[cccxxvii].

Another suggestion is that the universe is some form of self-learning AI that works out what laws are needed. With this idea, the universe sets out with some very basic rules, including ones for learning. It then acts like a neural network or super-brain, selecting the laws that allow the universe to evolve over time. It's not evolution by survival of the fittest (that's the multiverse option) but evolution by learning, adapting or mutating the universe's "DNA" to meet its needs. The universe may be smarter than we give it credit for[cccxxviii]. However, AI means "artificial" intelligence, and nothing's as "natural" as the universe. So, we're talking about some form of "intelligent" universe, which sounds a bit outlandish on the face of it.

When we look at matter, we find that any individual thing depends on something lower down the chain, such as molecules depending on the bonds between atoms, which depend on the electrons forming an atom's shell, etc. By following those dependencies back, you come down to about thirty fundamental constants of nature such as the strengths of gravity and the other fundamental forces and the mass of protons and electrons, and these kinds of values seem to be fine-tuned for the universe we have and the life that exists in it. Some of those constants aren't too critical, and changing the values won't cause much harm, but if you tweak others, atoms won't form, or the universe might collapse back on itself before stars and galaxies could come into existence. The "intelligent universe" idea has the universe manipulating its control panel to ensure that the right settings are there. Some people do question if some of these "constants" are really as fixed as we believe, or if they may have been changing as the universe grew[cccxxix]. So, maybe the universe has been tweaking the dial.

Looking at matter in finer and finer detail, we find it becoming increasing intangible as we go down through atoms which are made up of subatomic particles (protons, neutrons and electrons) with relatively vast volumes of empty space between them, and those particles can also be viewed as waves. Ultimately, we get down to the wavefunction of quantum theory. But that wavefunction is also what drives quantum computers, so might the universe be a gigantic quantum intelligence? Others see information as being fundamental, and the Integrated Information Theory/Hypothesis suggests that consciousness emerges in all sorts of systems that integrate information, and surely nothing integrates information as thoroughly as the universe does[cccxxx]. If the universe was a living being, it would at least solve the problem of how life got started on Earth, because Earth-life would just be an offshoot of universe-life, but then we'd have to wonder about how the intelligent universe got going at the Big Bang, and we're already wondering how and why the Big Bang happened.

But the intelligent universe idea is not the first suggestion about "intelligent" action from objects that we normally think of as basic matter. Gaia is the idea that the Earth itself is a complex, self-regulating system with its biosphere of atmosphere, oceans, etc., together seeking to create and sustain a suitable environment, much like some kind of intelligent neural network[cccxxxi]. I'm not sure if that is a good idea or not. Someone suggested a scenario where an AI was tasked with the job of finding and implementing a solution for the climate crisis, and it came up with the simple solution of eliminating all human life from the planet.

We think that life on Earth most likely started in the ocean, and there have been suggestions that the ocean itself should be seen as a single living creature, with the other creatures that live in it being part of the ocean-creature. That would be like the bacteria that forms our microbiome being effectively a part of us, in the same way that the fish and bacteria in the ocean are assisting in processing the material delivered to it by the streams and rivers. They haven't been successful at processing the plastic bottles that we dump in the oceans, but some bacteria are working on it. It seems that, a number of times in the past, the ocean has been overwhelmed and has belched up the hydrogen sulfide that collects on the ocean floor, leading to the anaerobic events that caused, or were part of, some of the mass extinctions. Hopefully we learn to treat it better before it decides it's had enough of us[cccxxxii], although the failure to pass the ocean protection treaty doesn't make that look promising.

The problem of where consciousness comes from has led to an idea that consciousness might actually be a fundamental aspect of all matter, so maybe there might be some slight chance that the universe is intelligent. Do you think I included enough "maybe's", "might's", and other "weasel words" there?

Chapter 15: Trauma, Tragedy and Triumph

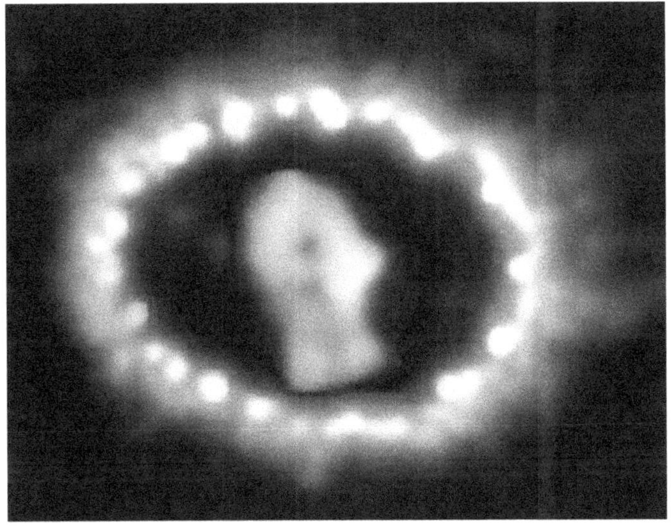

Big Bangs, Smaller Bangs, and Violent Disruptions

The universe has a rather dramatic way of growing and developing. It started with the biggest explosion ever, which led to the creation of the simple atoms. Those atoms combined and led to other massive explosions in the form of hypernovae, supernovae, and kilonovae, and those gave us all the more complex atoms and precious molecules that we have today. A supernova is about 10 to 100 times brighter than a kilonova, and a hypernova or collapsar is 10 to 100 times more powerful than a supernova[cccxxxiii].

Our Earth and other planets were built up by smaller planets and asteroids crashing into one another and combining. About the time that life got started on Earth it had to survive through the Late Heavy Bombardment that left as many craters per square mile on Earth as we see today on the Moon, but that bombardment might have brought some of the ingredients for life and created the locations where life could have started. As life developed, it came close to being wiped out many times, but always came back stronger and more resilient, and has now spread to just about everywhere on Earth, including places that seem, on the face of it, to be totally inhospitable. The universe loves big explosions and what we'd call catastrophic events, and that appears to be the way it grows.

To the End of the Earth

After the Sun exhausts the hydrogen in its core in about five billion years' time, it will have swollen into a red giant, engulfing the orbits of Mercury and probably Venus and, while Earth may escape being engulfed, it will be a charred, lifeless, magma-covered rock with its oceans and atmosphere blown away. 1.1 billion years from now (or earlier if we're not careful) the Sun will be 10% hotter than it is now and our oceans will start evaporating, which was what we imagine as being the first stage of Venus's transition into the scorching hellhole it is today. 10,000 years later, our oceans will be gone and ambient temperatures on the planet will be around 1,000°F (550°C), and still headed up. By 1.3 billion years from now, the temperature will have risen to around 3,600°F (2,000°C) and any alien visitor to the solar system will not see Earth as a place where they'd expect to find life.

If the Sun gets too close, friction with the Sun's outer atmosphere will slow Earth's orbital speed and it will spiral into the Sun. If not, as the Sun shrinks back, the Earth will find itself orbiting a hot white dwarf not much larger than itself. Over time, what's left of the Sun will cool and become a black dwarf, and Earth's orbit will decay due to gravitational wave emissions, and it will spiral into the dead star over a period of a trillion trillion years[cccxxxiv]. Disruption of the planets' orbits as the Sun becomes a red giant might end up throwing some, possibly including Earth, out of the solar system. A close encounter of our solar system with another star or black hole is another way that the solar system could be disrupted.

The Andromeda Galaxy is only a blur in the night sky, but it is a great spiral disk of about a trillion stars, plus a supermassive black hole larger than our Sagittarius A*, and it is all hurtling towards us at about 70 miles per second (110 kilometers per second). In about 4 billion years our galaxies will collide, stellar streams will get pulled out, and gas clouds will be colliding, sparking a minor explosion of star-births. As the two supermassive black holes spiral towards a collision, the gas around them will ignite and jets of intense radiation and high energy particles will bathe the combining galaxy and especially the central region of the new galaxy with a hot X-ray glow from matter falling into the newly combined super-supermassive black hole. Our scientists like to call the combined galaxy Milkomeda, but Andromeda is way bigger than us, and their scientists might have different views, so we might find it being called Andromaway. Our solar system will probably survive the merger, but the Sun will already be a red giant by then, so there will be no life on Earth to see the results. The galaxy merger will play out over billions of years, leaving a massive ellipsoidal collection of old and dying stars.

If the expansion of the universe continues, as it almost certainly will, the universe will continue to cool down, stars will fade away, black holes will probably evaporate, and all that will be left will be the occasional atom, and fluctuations in what is known as vacuum energy. But one of those fluctuations could, supposedly, result in another small and quiet Big Bang, and a new universe might spring into existence. All that we will need to do then will be to create a protected planet or space station to live on while the universe changes around us. Maybe turn it into a restaurant where we can relax as we watch the event. Someone has come up with a speculative idea for creating a bubble universe for us to reside in while the main universe transitions. We have time to work on the details, but there are plenty of other ideas for what might happen, not just the steady expansion and cooling. We'll look at those in the next section.

Part 4: Where Will It End?

Astronomical Alchemy

324

Chapter 16: Big Crunch or Heat Death?

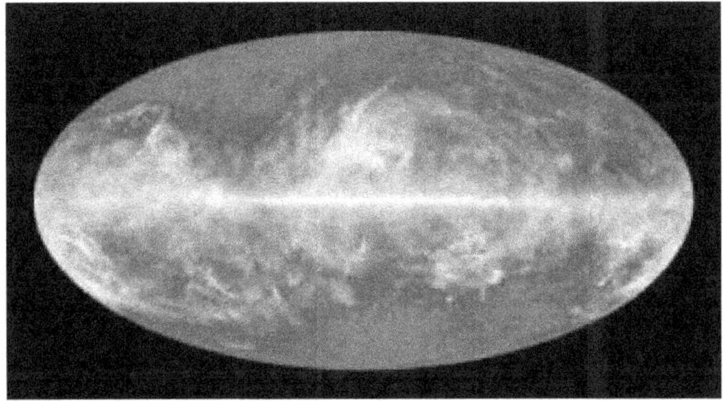

The Big Crunch

The cosmos appeared to be static when the Milky Way was assumed to be the entire universe, so it was thought that nothing much would change (Einstein's fudge factor – the cosmological constant - was added to his equations to keep the universe steady). Then other galaxies were observed, and Edwin Hubble showed that most galaxies were moving away from us, and suddenly we had the expanding universe. Playing that expansion backward showed that it must have started expanding around 13.8 billion years ago.

Looking to the future seemed to have options for the universe collapsing back on itself or else expanding forever, and that was thought to depend on the density of the universe, i.e., whether there was sufficient gravity to halt and reverse the expansion. At first it seemed that we were headed for the Big Crunch, and black holes were seen as a localized version of what the universe would become[cccxxxv]. That Big Crunch could reset the universe to its starting point, which led to the idea that there might be an infinite cycle of expansions and contractions, with the universe bouncing back and being reborn and then collapsing and dying over and over.

So, let's look at the idea of the expansion slowing down and stopping, then reversing direction to end up in a Big Crunch. Gravity is the weakest of the four forces, but its reach is infinite so it's natural to expect the expansion to be slowing and then to have it start reversing its direction. Because of the time it takes light to reach us, distant galaxies might have already stopped and reversed direction, but we wouldn't see that for a long time. We would first see nearby galaxies stopping their movement away from us and then start heading inward like Andromeda is doing, and over time we'd see those further out doing the same. The earliest that one might anticipate a Big Crunch occurring would place it as billions of years out, with the universe being now in its middle age.

As the collapse occurred and galaxies got closer, galaxy mergers would become more frequent giving more starbursts and radiation, but the afterglow of the Big Bang is also being compressed and blue-shifted to uncomfortable levels, and planets start to burn up. All the radiation from every star and black hole that has existed also gets compressed, resulting in so much concentrated radiation that it ignites the surface of stars, ripping them apart and filling space with hot plasma. So, maybe the end of the universe isn't such a good place for a restaurant after all, although there would be plenty of heat for cooking things.

One interesting idea is that the universe could bounce back into a new one, but how that physics would work is unclear. From the kind of things that we see happening with black holes, compressing mass together leads to greater gravitational pull until nothing can escape, let alone bounce back as a new universe. In terms of the radiation field, it seems that the universe would get messier with every cycle, making the recycling of matter and energy not so neat, and that could lead to any following iteration of the universe-cycle not being as conducive to life as the previous one.

Anyway, a crunch depends on the amount of matter in the universe. Counting the visible matter showed us that we were missing something, and that led to the discovery of dark matter. We haven't detected what dark matter is made of, but we can see its gravitational effects. With the visible baryonic matter and the dark matter, it was still hard to say which side of the "critical density" we were on, although it appeared to be on the side of continuous but slowing expansion. Efforts in the 1990s to measure how fast the expansion was slowing down gave the surprising result that expansion was speeding up, and the term "dark energy" entered public awareness. A universe whose expansion is accelerating is one in which the influence exerted by the things in it is shrinking. That doesn't mean a crunch isn't possible if the expansion starts slowing again, but it's making it look unlikely. On the other hand, some think that dark energy may be a scalar field, like the inflaton field that supposedly drove inflation, and that field is assumed to have collapsed and transformed itself into matter. If dark energy collapsed, stopped pushing the universe apart, and converted itself into matter, that would mean more matter and more gravity to slow things down and a Big Crunch is back in consideration again.

Heat Death or Big Freeze

For now, it seems as if the universe continues to expand, galaxy growth and star formation ceases, stars fade away, black holes absorb the remains of stars and planets and useful heat disappears from the universe, giving us what is called Heat Death or a Big Freeze, with the universe getting colder and darker. It's called "Heat Death" in relation to heat being "disordered motion of particles or energy", and it is more the death of "useful heat" rather than death by being roasted. To be useful as a source of energy, heat needs to flow from a high point to a low one, but we are considering a time when heat is spread out evenly and the constant expansion of the universe is spreading it thinner, making the effective temperature colder and colder and the likelihood of there being useful heat becomes vanishingly slim.

The current view is that normal matter makes up 5% of the universe, dark matter is 27%, and the remaining 68% is accounted for by dark energy. Dark energy is sometimes seen as Einstein's cosmological constant, but that it is now used to push things apart rather than holding them steady. With the growing expansion caused by dark energy, all but the closest galaxies are expected to become too far away and traveling too fast for us to see in a couple of trillion years. If the expansion doesn't fluctuate with time, then it will be carrying other galaxies away and leaving us alone in our visible universe. If dark energy is a constant, then it may push galaxies apart, but it won't affect the galaxies themselves. Super-galaxies will end up isolated with no fresh supply of gas coming their way, so stars start to die out, and black holes grow but then they'll start to evaporate away. Red dwarfs will become the only remaining stars and will be the last hangouts for any surviving lifeforms in the universe.

Maybe 100 trillion years in the future, star formation totally ceases as white dwarfs, neutron stars and black holes absorb any remaining gas. About a googol years from now (a googol is a 1 followed by 100 zeros or 10^{100}), supermassive black holes will have evaporated by Hawking radiation, and we enter the Dark Era with no noticeable matter. Entropy will max out as all of the energy (heat) gets uniformly distributed, and everything settles down at a temperature of just above absolute zero.

Neutrons decay eventually into a proton, an electron, and an antineutrino, and protons are believed to decay over a period of around 10^{33} years, although we've never seen that happen so far. Universal expansion continues to accelerate, stars have burned out and particles are decaying, the last of the black holes evaporate and explode, and dark energy is all that's left. We're not sure exactly what happens to the dark matter, but presumably it decays as well. Energy gradients can't exist, so heat is useless.

However, on the small quantum scale, and larger scales if you wait long enough, unpredictable fluctuations could spontaneously shift some part of the system into a lower-entropy state at random. Statistical mechanics has an idea called the Poincare recurrence, where given enough time any possible state will recur. So, a new Big Bang might suddenly occur, a galaxy might instantly appear, or just a solar system (because it's simpler than a whole galaxy), or simpler still is a Boltzmann Brain that might come complete with memories of a universe. We will be long gone by that time, anyway[cccxxxvi].

The Big Rip

Dark energy has negative pressure. Pressure is another kind of energy which is equivalent to mass and, according to general relativity, it is gravitationally attractive, the reverse of dark energy. Negative pressure can effectively cancel out the mass as far as creating gravitational attraction is concerned. The relationship between a substance's density and its pressure uses a number called "the equation of state parameter", ω, which is the pressure divided by the energy density. For dark energy, if $\omega = -1$ exactly, then the pressure and density are exactly opposite, and dark energy is a cosmological constant. It will have a total energy density that is exactly constant over time as the universe expands.

However, if ω is not exactly equal to -1, there's the possibility of a "big rip" scenario in which dark energy gets stronger, expansion accelerates, and even gravity-bound galaxies get pulled apart. To determine if that is likely to happen, we need to know the nature of dark energy.

Dark energy appears to exist everywhere, woven into the fabric of space, with its only effect being to stretch space out gradually. That means that as the universe expands, the amount of dark energy has to increase to keep the density constant. But if dark energy is more powerful than that, we can get what is called "phantom dark energy", and it has the prefix "phantom" because it would involve physics in which we would have energy flowing faster than light. In that case, the dark energy will tear the universe apart in a calculable time, and this is what is known as the Big Rip. Some of the explanations for the differences in the measurements of cosmic expansion involve dark energy of the kind that would produce a Big Rip.

If a Big Rip is going to happen, then it could occur at some time over 100 billion years in the future, but it is most commonly expected to occur something over 188 billion years from now. About two billion years before the final rip, galaxy clusters would get torn apart, 140 million years before the rip we'd see the Milky Way and other galaxies being destroyed, 7 months before the rip we'd see the solar system would fly apart, 1 hour before the end we find the Earth would start disintegrating, and 10^{-19} seconds before the Big Rip we find that atoms come apart.

Universal Transformation

There is another idea, which sounds almost as bad as the Big Rip, and maybe worse because it could happen at any time. That idea says that our universe might perform a quantum trick called tunneling and transform itself into a universe with different properties, changing the laws of physics and the constants of nature on which our atoms and ourselves depend.

The laws of physics are energy-dependent, establishing how particles and forces interact, and the Higgs field, like any field, will try to find its lowest energy state. In its current state, it might not be at the real bottom, but just be sitting in a dip of the Potential/Field Value curve. That would mean that the Higgs vacuum is not completely stable, only stable ... ish for now. It could transition from its current value (called the false vacuum) to the other state (the true vacuum) because of some high-level event (e.g., an ultra-high-energy explosion or the catastrophic final evaporation of a black hole) or via quantum tunneling. There are some indications that the Higgs field is clinging to such a "false vacuum", and any transition might create a tiny bubble of "true vacuum" where the processes of physics follow different laws, and that bubble could start expanding. Anything in the bubble's path would be hit by the intensely energetic bubble wall traveling at the speed of light, destroying any matter in its way and you'd never see it coming. The only relatively safe spots would be those regions of space that are a long way outside of this expanding bubble where the universe's expansion has resulted in things moving away from the bubble at faster than the speed of light. But inside the bubble there might be new physics creating new kinds of matter. There could be multiple such bubbles with new matter in them creating their own universes and expanding outward at the speed of light. But there would be parts of the universe outside of these bubbles that could be expanding faster than light, so this additional form of multiverse should not have to worry too much about one universe crashing into others too often.

Happily, vacuum decay is very unlikely to happen anytime in the next many many many trillions of years. The kind of high-energy event needed to trigger it would have to be much more energetic than the most devastating explosions we've witnessed in the cosmos, so high it's hardly worth worrying about.

Or is it? There is another idea that, as two black holes collide, the extreme gravity between them would cause a foam of small bubbles of true vacuum. Happily, it is expected that these bubbles would then be drawn into the black hole and safely lost to our universe. However, such bubbles do expand rapidly and, theoretically, could escape from being drawn into the black hole. That said, black holes have been colliding since early in the life of the universe and we are still here, so the danger appears to be very small[cccxxxvii].

With quantum mechanics, a particle or field might, very

rarely, sail right through solid objects in a process known as quantum tunneling. How it gets there is unclear, but they obey the mathematics of waves that are spread out through space, and we use this kind of tunneling with some of our technology, but we're doing nothing that affects the Higgs field itself. Creating Higgs bosons is different from messing with the actual field. Nevertheless, the Higgs could, in theory, tunnel by itself to a lower energy level. The best estimates of the probability of such tunneling happening currently suggest that it is not likely to happen for more than 10^{100} years in the future.

If the Higgs field really is only metastable, that might contradict the idea of inflation, because quantum fluctuations during inflation seemingly should have been able to trigger vacuum decay, destroying the universe before it got started. So, since we are here, it seems that the universe is more stable than some fear it to be. Alternatively, we can imagine that inflation didn't occur, in which case universal transformation becomes a possibility again. Maybe I'm getting to like the idea of inflation a bit more ...

The Big Splat

One idea for why gravity is so weak is that it's leaking into another dimension. In the "large extra dimensions" scenario, there is one or more extra dimensions that we can't access. The space part of our spacetime, in this scenario, is limited to a 3-D brane (membrane) and a larger space is imagined extending outside of it in some new direction or directions. We've stepped down from astronomical scales to those of particle physics, so "large" in this context might mean that these extra dimensions extend no more than a millimeter in the new directions. In this hypothesis, the forces except for gravity are confined to the brane, and to those forces the larger, higher-dimensional bulk doesn't appear to exist. But gravity is supposedly able to act on our spacetime as well as that outside our 3-D brane. So, some of the gravity from a massive object in our space loses a bit of its strength by leaking out into the unseen bulk. Scientists are trying to measure gravitational changes on the millimeter and lower scales, to try to observe the amount of gravity we are losing but, so far, we haven't found any evidence that it's happening. However, if such branes do exist, then interactions between branes could lead to our destruction.

The idea here is that the Big Bang resulted from two branes (one of which was ours) bumping into each other, and that collision was the cause of a Big Bang event in both branes. If we happened to bump into another brane, that could wipe us out. There could be a Big Splat leading to a new Big Bang occurring.

Some people suggest that data from black hole collisions contradicts these theories involving gravity leaking into a higher dimensional void, so we may be able to avoid that ultimate cosmic smash-up.

Conformal Cyclic Cosmology

Roger Penrose proposed a cyclic cosmos in which our

Big Bang was born from a previous cycle's Heat Death with the universe cycling through periods of expansion and contraction. This Conformal Cyclic Cosmology suggests that entropy works differently near a singularity, implying that entropy would be very low at the boundary between cycles and doesn't need inflation. If this is true, then some imprint of past cycles might show up as features in the CMB because electromagnetic radiation (such as we could expect from the Hawking radiation from evaporating black holes) could pass from one iteration of the universe to the next[cccxxxviii].

Poof!

One final thought on the subject: It has been pointed out that if the universe is infinite, and especially if it would go on for infinite time, then anything that could conceivably happen, can and will happen. One of the things that could happen is that the universe could suddenly end. After all, we think the universe suddenly came into existence, so it could just as easily go out of existence. You'd be going about your business and, just like that, everything's gone. Of course, if that happened, then the universe would have a beginning and an end, so there wouldn't be infinite time for everything to happen in, so it probably wouldn't happen. Then there would be infinite time, so …

Chapter 17: Another Transfiguration?

States of Matter

Matter is normally thought of as being either a solid (having a fixed shape with atoms in a rigid three-dimensional lattice), a liquid (which flows to fit the shape of a container and with which the atoms are free to move around each other), or as a gas (in which case it expands its shape to fit its container, and there are energetic atoms flying all over the place, barely touching one another). To move from one state to the other, you add or remove energy (i.e., heat). However, these categories don't always work the way we'd like them to. Glass retains its shape even though its atoms are disorganized like a liquid. Liquid crystal (used in some smartphone displays, etc.) can flow like a liquid although its atoms are arranged like a solid.

Plasma (that we come across a lot in space) is often called the fourth state of matter, although perhaps it should be the first, because about 99% of ordinary matter in the universe is in the form of plasma. It is a gas that has had its atoms split apart into charged particles, so plasmas conduct electricity although, shape-wise, they are like a gas.

Matter can get squeezed so much that the Pauli exclusion principle comes to rule, where some types of identical particles can't occupy the same space and be in the same quantum state. Consequently, in white dwarfs the matter may technically be a gas or plasma, but it becomes virtually incompressible (degenerate matter). At higher pressures, quarks hit the Pauli principle and theoretically form quark matter, which could be the densest stuff outside of black holes.

Bose-Einstein Condensate

A frozen cloud of molecules that share the same quantum state and behaves like a single entity has been created. This was a Bose-Einstein Condensate (BEC) made up of thousands of cesium molecules that were cooled to near absolute zero using lasers to remove their energy. BECs are often called the fifth state of matter (after solids, liquids, gases, and plasmas) and their particles share the same quantum properties as each other, all working together in the same way, as if they were one giant molecule. BECs have been created with atoms since the 1990s but putting molecules in that state proved hard. In the experiment that we are discussing they started with a single layer of atoms in a BEC and used a magnetic field to induce pairs of atoms to form molecules while still in the BEC state. They remained stable at 10 nanokelvins, just above absolute zero[cccxxxix].

Superconductors, Superfluids and Supersolids

A normal conductor of electricity will become more conductive (i.e., lose more of its electrical resistance) as it cools, but a superconductor will go through a phase transition at a "critical" temperature and suddenly lose all of its electrical resistance[cccxl]. Superconductors act as if they are a single super-particle, along with conducting electricity with zero resistance. That normally only occurs at super low temperatures of about -459.4°F (-273°C or 0.15 K), although we now have ones that work at higher temperatures. Some materials have been found to become superconductive around 90K (-298°F or -183°C) which can be achieved using liquid nitrogen for cooling. These are called high-temperature superconductors[cccxli], with what is classed as "high-temperature" starting at 30K (-406°F or -243°C), so "high" is relative.

A magnet moving close to a superconductor will induce electrical currents flowing on the material's surface, creating a magnetic field that is opposite to the field of the magnet. In that way, you can levitate a magnet above a superconductor. Superconducting magnets are used in MRI machines, creating that constant hammering noise while the medical technician is telling you to relax.

At these low temperatures needed for superconductors, pairs of electrons can become bound together as "cooperpairs" through the exchange of phonons, and then they act like bosons. That means that, unlike single electrons, these cooper pairs can overlap, with the result that they appear together like a single super-particle. Now it can move as one while showing no resistance.

If you try cooling helium and some other substances to just above absolute zero, instead of turning into a solid, the substance will become a superfluid, allowing it to flow uphill and, if stirred, it will never stop rotating. Superfluid behavior also occurs in Bose-Einstein condensates where a supercooled cloud of atoms behaves as a ghostly fluid. This state is thought to be similar to the quantum goings-on occurring near the edge of a black hole and superfluids have also been suggested as a basis for spacetime itself and have been called upon to explain some of the mysteries of dark matter.

Supersolids derive from a theory that suggests that holes in a solid lattice of atoms can, at very low temperatures, form a kind of ghostly matter that can pass through other solids, but it is still only theoretical and has never been observed.

Absolute Zero Universe

General relativity breaks down at the small scales near singularities and has problems at the very large scale (where dark energy takes over). As dark energy pushes the universe apart, we have large voids that look stable, with the same level of energy everywhere. Matter does decay but, overall, it retains the same net mass although more spread out, but the proportion of dark energy increases as the universe expands while the dark energy "density" stays the same throughout. The result of all that is that it seems that the universe could be heading towards a stable state that's almost empty and almost at absolute zero.

The massive voids in the universe are what is growing due to universal expansion, but what if it is the voids that really count? We looked at the isotropic expansion of the universe as being a bit like a 4-D bubble, with the galaxies that we see being a 3-D bubble skin. But the "air" in a bubble is what makes it what it is and is all that's left when the bubble bursts. The growing voids get very cold and, as we've seen, some very interesting things happen near absolute zero, looking rather like perpetual motion. We've seen that matter behaves very differently near absolute zero and we've seen that molecules can be formed in a BEC, so we could have different types of matter forming in the void. Quantum computers need such low temperatures to operate, so maybe the universe could become an intelligent quantum computer, if it isn't already.

Of course, the perpetual motion that we see in superconductors and the like doesn't mean perpetual useful energy. Start using that energy and you are continuing to increase entropy, but the universe could prolong its useful life and who knows what other tricks is might have up its cosmic sleeves.

Appendix A – Sky Watching

The universe has had a long and fascinating history and still has an exciting future ahead of it, and in the meantime it provides us with a lot of wonders to observe in the present.

The Sky in Motion

As you gaze at the night sky, the stars will be rising from the east, making their way across the sky, and setting in the west as the Earth rotates from west to east beneath the sky. If you observe the sky at around the same time each night, you'll find that a particular star rises a bit earlier each day, and that is due to the Earth's motion in its orbit around the Sun. So, if you are looking for the same star or constellation at the same time each night, as the days pass you will find it further and further to the west, until it has disappeared below the western horizon. One complete orbit of the Sun gives us one year (just over 365 days), so the shift is about one degree per day[cccxlii].

Assuming that you have a mostly unobstructed view to the horizon, and that the sky is dark and not cloudy, you can see about half of the celestial sphere at any one time. However, almost the entire sky becomes viewable throughout the night. Immediately after sunset, stars that are a bit to the east of the Sun are visible in the western sky, and before sunrise those stars that are a bit west of the Sun become visible in the east, and throughout the night, all the stars and constellations in between those ones will pass overhead.

You will often (including in this book) see stars and constellations described as being those visible in spring, summer, autumn or winter, and what that is referring to are those stars that are nearly overhead around nine or ten p.m. at those times of the year. The point directly overhead at any time you are out looking at the stars is called the zenith, and the meridian is the great circle running through that point and the two celestial poles (which we'll get to shortly). If you stay up all night, you can see almost all the visible stars and constellations. Okay, that is true, but only if you are viewing from the equator. Since you are more likely to be somewhere north or south of that, there will always be some stars that remain hidden below the horizon, just as there will be circumpolar stars that never drop below the horizon.

Orienting Yourself in the Night Sky

The Celestial Poles are the points in the sky directly above the North and South Terrestrial Poles, and the entire sky appears to rotate continuously about these Celestial Poles due to the Earth's rotation on its axis. The North Celestial Pole is currently near the star named Polaris, commonly called the North Star or Pole Star. For our friends in the southern hemisphere, there unfortunately isn't a visible star that can currently define the south celestial pole.

The Celestial Equator is the imaginary "great circle" in the sky formed by extending the Earth's Equator out into space, so it is always exactly halfway between the Celestial Poles. The "Right Ascension/Declination" coordinate system is based on the Celestial Poles and the Celestial Equator and is analogous to the terrestrial "Longitude/Latitude" system. To define the position of a star in the sky, you can first see what angle it makes with the celestial equator, which itself would be zero degrees, so the north celestial pole is $+90°$ and the south celestial pole is $-90°$. That would be equivalent to finding its latitude but, in this case, it's called the Declination. Then, imagine lines of longitude forming half of a great circle running from one celestial pole to the other, crossing the celestial equator at some point, and one of these "lines of longitude" is running through the star we are trying to locate. The point where that great semi-circle crosses the celestial equator defines the equivalent of the longitude of the star, but that point is normally defined in terms of time (hours, minutes, and seconds) from the point in the sky defined by the Vernal (or Spring) Equinox. Since there are 24 hours in a day, halfway around would be defined as 12 hours of Right Ascension. We'll get to the equinoxes and solstices, but the Spring or Vernal Equinox is known as zero degrees or 0 Hours of Right Ascension, the Summer Solstice is $90°$ or 6 hours of Right Ascension, the Autumn Equinox is $180°$ or 12 hours of Right Ascension, and the Winter Solstice is $270°$ or 18 hours of Right Ascension from the Vernal Equinox.

The Galactic Center is located in the summer Milky Way within the constellation Sagittarius near the border with Scorpius. That is how the supermassive black hole at the center of our galaxy comes to be called Sagittarius A* (pronounced Sagittarius A-star). The Galactic Anti-Center (the point in the sky directly opposite the Galactic Center from our viewpoint) is found in the winter Milky Way near the star Al Nath at the border of Taurus and Auriga.

The Ecliptic is another "great circle" in the sky, and it denotes the path that the Sun traces across the sky during the year due to Earth's motion in its orbit. In effect, the Ecliptic marks the plane of the solar system, at least as viewed from Earth. The orbital planes of the planets and the Moon all lie near enough along that plane, with the result that the planets and the Moon are found on or close to the Ecliptic. A line perpendicular to the ecliptic plane determines the locations of the Ecliptic Poles, which are about 23.44 degrees off from the Celestial Poles (equal to the Earth's tilt, of course). The North Ecliptic Pole is in the constellation Draco, about halfway between Altais and Aldhidah. We can consider the Ecliptic plane and poles as "fixed" in the sky, although nothing is really fixed in the universe. But these poles are more fixed than the Celestial Poles. That is because the Earth "wobbles" a bit with its spin, making the Celestial Poles move in relation to the background stars, while our orbit stays basically the same and the Ecliptic and its poles rely mostly on our orbital path.

With the Ecliptic being the path that the Sun takes

across the sky throughout the year and the Celestial Equator lying directly over Earth's equator, it follows that these two great circles will cross each other at the point where the Sun is located as it passes directly over Earth's equator, and that event occurs twice a year. These are the Equinoxes, meaning "equal night" since day and night are of equal length on these days. The bending of light by our atmosphere means that we can see over the horizon a bit, with the result that we see the Sun rising about two minutes earlier than it "really" does, and setting later, so daytime is actually slightly longer than night on the equinoxes.

The Vernal Equinox is known as the "Ascending Node" since the Sun is "ascending" from the south to the north. The Autumnal Equinox is the "Descending Node" marking the first day of autumn, around September 21. The term "Equinox" relates to two specific points in the sky and also to the events of the Sun passing through these points, so Equinox can mean a location in the sky or a date and time. When astronomers use the term, they are usually referring to the specific points in the sky.

As the Sun "ascends" into the Northern Celestial Hemisphere after the Vernal Equinox, it eventually reaches a point of maximum northern movement, which denotes the Summer Solstice. The first day of summer in the Northern Hemisphere occurs when the Sun reaches this point, around June 21. It is known as "The First Point in Cancer", and hence we have the Tropic of Cancer where the Sun is overhead on that day although currently the Sun is located in the constellation Taurus at that point in time. The Sun then moves southward until it reaches the Autumnal Equinox as it crosses the equator and enters the Southern Celestial Hemisphere. The autumnal equinox marks the "first point in Libra". The Sun continues to "descend" until it reaches a point of maximum southern declination at the Winter Solstice. The first day of winter in the Northern Hemisphere occurs when the Sun touches this point, around December 21, which is known as "The First Point in Capricorn" although currently located in Sagittarius.

Because the Celestial Poles and Celestial Equator are always slowly shifting, and because right ascension and declination are based on these and on the location of the vernal equinox, the Celestial Coordinate System is also constantly shifting. When professional astronomers use this system, they calculate the Right Ascension/Declination coordinates accurately to the exact moment in time, called the "Epoch" of the coordinates. However, in 2000 they came up with ICRS (the International Celestial Reference System) that identifies locations in relation to 212 celestial radio sources, so they no longer have to be concerned about shifting equinoxes[cccxliii]. For amateur astronomers and for common reference purposes it is customary to specify the right ascension/declination coordinates for objects at the nearest fixed 50-year interval Epoch. When this book was published, we were using the Year 2000 Epoch coordinates.

The Zodiac is a 30-degree wide band, centered on the Ecliptic, and the Sun, Moon and planets all travel within this band. The Ecliptic passes through the twelve zodiacal constellations plus Ophiuchus, which is sometimes called the thirteenth constellation of the zodiac. This Zodiacal band is divided by western astrologers into 12 equal sections, each 30 degrees across, which are called "Houses" and, probably unfortunately, they are given the same names as for the zodiacal constellations.

The astrological year starts at the spring (or vernal) equinox in March, when the Sun crosses the celestial equator as it moves from the southern hemisphere into the north. That is called the first point in Aries. However, because of the Earth's wobble, the vernal equinox has moved from where it used to be, and the Sun is now in the constellation Pisces at that point in time. Nevertheless, astrologers would still place you in Aries if you were born at the end of March – they normally go by the "houses" not the physical constellations that the Sun is in. If you wondered why more convenient dates for birth signs weren't chosen – now you know.

So, in astrology, your birth "sign" really defines your birth "house", not the constellation. The "houses" might be named the same as twelve of the constellations, but they need not correspond with them as places in the sky. Instead, they are tied to the equinoxes, and we have seen that the equinoxes and solstices move over time.

Telescopes

Refracting telescopes use two lenses, and this type of telescope was discovered accidentally by an optician, or rather by a couple of his customers. A large convex lens at the front collects the light and sends it to a smaller concave lens that serves as the telescope eyepiece. Galileo introduced this type of telescope to astronomy and used it to revolutionize our understanding of the solar system.

A reflecting telescope uses a concave mirror to focus incoming light and it is the most commonly used type of telescope for astronomy now. That is because it is easier to make a large mirror than a large lens, and a larger collector of light lets you see more detail in an object or see further and therefore see more stars and galaxies. Newton is seen as the inventor of the reflecting telescope. Reflecting and refracting telescopes are both optical telescopes collecting photons that fall in the visible range.

(1) is where the finderscope would be, had I remembered to attach it, and that is very important for aiming the telescope in the right direction, and it may use a laser to help to aim directly at a star or planet. (2) is the tube or body of the telescope itself, at the bottom of which is the

primary mirror that collects the light from the stars and sends it to another mirror that reflects it to the eyepiece (3). (4) is the tripod supporting the telescope, and each leg usually has an extension that is held tight by the leg clamp. (5) is a brace that holds the legs steady and can also support an accessory tray. (6) is the mount that holds the telescope to the tripod.

There are other types of telescopes that pick-up regions of the electromagnetic spectrum outside of the visible range. These other types include radio telescopes, X-ray telescopes, and infrared telescopes. They might use a similar sort of design to a reflecting telescope for focusing the signal, because they are all electromagnetic waves, just as visible light is. If you do an Internet search, you will find ways to convert an unwanted satellite receiver dish into a simple radio telescope. Then we have gravitational wave and neutrino "telescopes", which are very different, and definitely not something you'll be using in your backyard.

History of Constellations

People are good at seeing patterns in things, and the patterns that people saw in the stars became the constellations we talk about. We know that many ancient civilizations saw patterns in the stars, and they associated these patterns with particular aspects of nature or with their myths, and many of the constellations we know today have come down to us from Babylonian times. One of the best-known constellations is the Plough or Big Dipper, except that that is not actually a constellation. The Plough (for UK readers) or the Big Dipper (for US readers) is part of the constellation Ursa Major, or the Great Bear. The Big Dipper is known as an asterism and consists of the seven brightest stars in the constellation. In other parts of the world, it has been seen as an axe or an oxcart, and similarly shaped objects.

The Babylonian astronomers produced the earliest catalogs of stars and constellations that we know of, although there are indications that they may have inherited some of these from the earlier Sumerians. Nevertheless, we can say that the majority of what we now call constellations date back at least to the time of the Babylonians. The Greeks and Romans basically adopted the Babylonian system and attached some of their legends and heroes to them, and most of the myths and legends that are associated with the constellations come from the Greeks. To them, the constellations were visualized as being semi-divine spirits that strode across the heavens.

Today we have 88 official constellations, mainly derived from the Babylonian system, with some others added to fill in gaps and to include the portion of the southern sky that was not visible from the northern hemisphere. Technically, to professional astronomers a constellation is a particular region of the sky, and the 88 constellations cover every square millimeter of the celestial sphere. All of the stars within one of these regions are seen as being part of that constellation. Any grouping of stars within a constellation that is considered as forming a shape is called an asterism, so Ursa Major or Big Bear has at least two asterisms: the grouping that represents the Big Bear itself, and the seven brightest stars from that grouping that we call the Plough or Big Dipper. In this book, I'll normally refer to the main asterism as simply being "the constellation" and reserve the term "asterism" for other groupings. Such asterisms sometimes spread across more than one constellation, like the Winter Hexagon and Summer Triangle that help us find our way around the night sky.

Other societies, outside of the Babylonians, Greeks and Romans, have also come up with their own interpretations of the star patterns.

The Chinese divide the northern sky into five enclosures, and they have twenty-eight mansions along the ecliptic which are divided up into Four Symbols. But there are similarities with the Babylonian system, leading some people to believe they had a connection in the past. The Hindus also have 27 or 28 sectors (known as mansions or nakshatra) along the ecliptic in their interpretation of the night sky.

Other civilizations, such as the Celts of Ireland and Scotland, came up with their own constellations and associated legends, but the influence of the Roman Empire led to many of these being replaced by the Babylonian/Greek/Roman constellations and their associated legends.

The Seasonal Skies

Even though the sky changes throughout the night, we associate the "evening" sky (that visible almost overhead about 9 or 10 p.m. local time) with the time of year in which we view it, since this is when we are more likely to be awake and noticing the night sky. It is also often warmer and hopefully drier than during the midnight and morning hours, and you want to be comfortable while enjoying the view. Consequently, the night sky is often divided into four seasonal quadrants based on what is overhead in the evening sky of each season. The north and south circumpolar constellations (those that remain mostly visible throughout the year at mid northern and southern latitudes) are often kept separate.

Northern Circumpolar Constellations

Starting with the constellations that are visible year-round (at least for those of us in the northern hemisphere), the easiest asterism (part of a constellation) to find is the Plough, also known as the Big Dipper, which is part of the constellation Ursa Major. If you look at the two stars at the end of the "bowl" of the Big Dipper and use them as pointers "upwards" they will take you to Polaris, the North Star, at the end of the "handle" of the Little Dipper (Ursa Minor). If you continue that line through the sky, it will take you past the tip of the constellation Cepheus and on to one end of the W or M (depending on time of year) that is the constellation Cassiopeia. Once you have located the Big and Little Dippers, you should be able to locate Draco, which snakes between them, although its stars are a bit faint.

For each of the seasonal skies, I give a summary of the

myths and legends associated with the main historical constellations. You will very likely come across variations of these legends – stories often get changed with the retelling and there have been millennia for variations to accumulate. I'm just giving a short version that might be of interest to you.

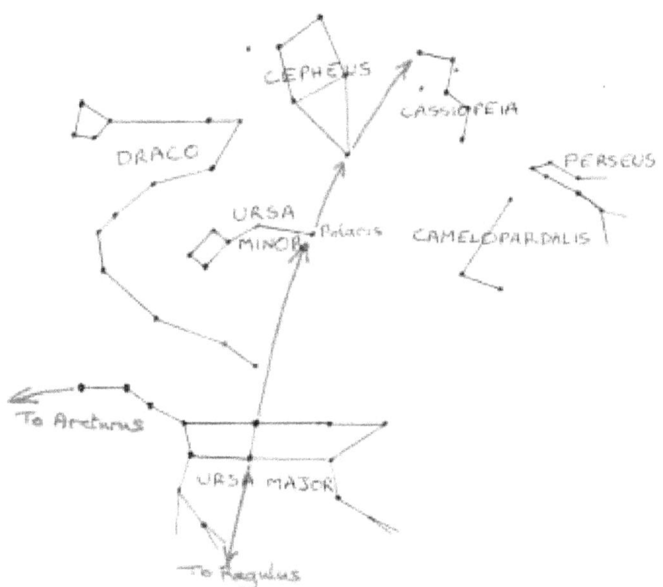

Ursa Major & Ursa Minor: Zeus (or Jupiter as the Romans called him) fell in love with the beautiful wood-nymph Callisto, a huntress. Hera (or Juno to the Romans), Zeus's wife, wasn't happy when she heard that Callisto had given birth to a son named Arcas, so she turned Callisto into a bear. When Callisto (as a bear) saw her son in the woods, she approached him but could only growl. Thinking he was being attacked, Arcas went to kill the bear with his spear, but to protect Callisto, Zeus changed Arcas into a bear as well and tossed them both by their tails into the northern sky. That is supposed to account for both of them having unusually long tails for bears.

Draco: The goddess Athena was attacked by a dragon during the Titan war, but she caught it in her hands and flung it into the sky, wrapping it around the north celestial pole.

Cepheus, Cassiopeia, Perseus, Andromeda: (OK, most of Perseus does not fall into the part of the sky that we consider to be north circumpolar, and Andromeda definitely doesn't, but the Greeks linked them together in their myths, and this is my book and my rules, so I will link them too. Live with it ;-) Cepheus was king of Ethiopia, Cassiopeia was his beautiful and vain wife, and Andromeda was their daughter. Cepheus was also one of Jason's Argonauts on the quest for the Golden Fleece. Cassiopeia claimed that she and her daughter were more beautiful than Juno and the Nereids (sea nymphs), so the Nereids got Poseidon (also known as Neptune, the sea god) to send a flood and a sea monster to destroy Cepheus's land. Andromeda was chained to the cliff as a sacrifice to the sea monster in order to save the land. Perseus saw her as he was flying home with the head of Medusa and fell in love with Andromeda and saved her by killing the monster with the magical sword of Hermes and/or the head of Medusa. Perseus is best known for beheading the Gorgon known as Medusa (which he achieved without being turned to stone by using his shield as a mirror and so never looking directly at her). His main purpose in getting Medusa's head was to save his mother, Danae, from a lecherous king (Polydectes).

Spring Constellations

In spring, if you find the Big Dipper and use the same two stars at the end of its bowl, but then imagine a line coming from them away from the Pole Star, that line will point in the direction of Regulus in the constellation Leo. Castor and Pollux, in Gemini, are both fairly bright stars and you can use them as pointers in a southerly direction, first curving down to Procyon in Canis Minor, and then continuing on to brilliant Sirius (the Dog Star) in Canis Major.

Caster is actually a six-star system; Pollux is an orange-hued giant star with at least one known planet; Procyon is a white main-sequence star with a faint white dwarf companion; Sirius is another white main-sequence star with a faint white dwarf companion. Regulus is another multi-star system consisting of four stars.

Gemini: Although they were twins, Caster was mortal, since his father was the human Tyndareus, while Pollux was immortal because his father was Zeus. Zeus got around a bit. The two boys were firm friends, and when Castor died at the Olympic games, Pollux asked Zeus to let him die too, and Zeus placed the two of them together in the sky as Gemini (the twins).

Cancer: Cancer represents a crab that was sent by Juno to bite Hercules's heels as he battled the Lernaean named Hydra. Hercules crushed the crab, but Juno was so pleased with its efforts that she placed it in the sky.

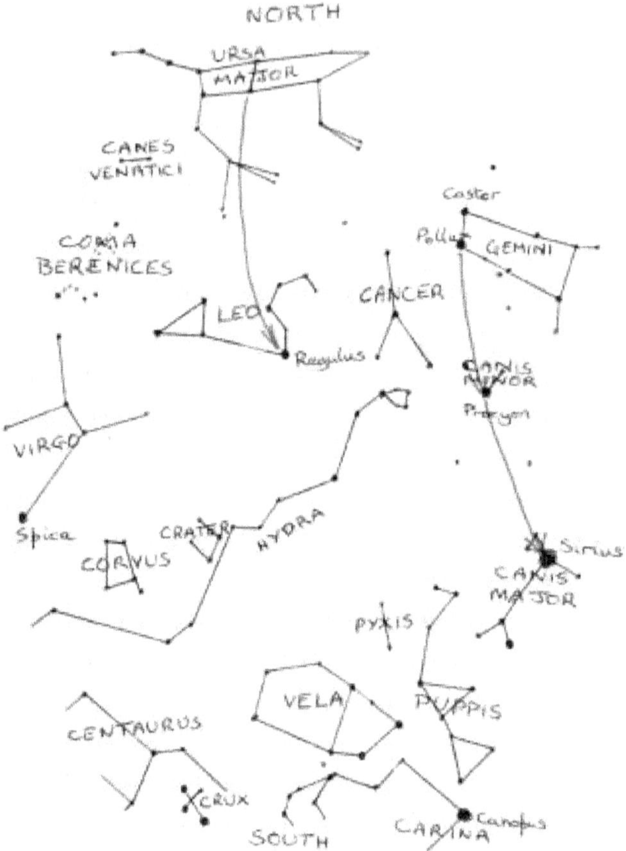

Leo: This lion had hide that was impenetrable to weapons and it terrorized the surrounding countryside. Hercules was commanded to kill it and that became the first of his twelve labors. He succeeded by strangling the animal and he wore its skin as a trophy. Hera placed the lion in the heavens.

Crater, Corvus, and Hydra: Apollo sent a crow to fetch

some water, but it wasn't diligent and spent some time resting. After finally getting the water in a cup, the crow took a water snake back as well as an excuse for its tardiness. Apollo wasn't taken in by the story and angrily tossed the crow (Corvus), the cup (Crater), and the water snake (Hydra) into the sky.

Summer Constellations

The handiest asterism in the summer sky is called the Summer Triangle, which spans three constellations. It actually shows up better in the image for the Autumn Constellations (in summer you would also be able to see most of the spring and autumn constellations depending on the time of night). The Summer Triangle consists of the brightest stars in Cygnus (Deneb), Lyra (Vega), and Aquila (Altair). If you continue the line from Deneb to Vega, it will take you across Ophiuchus and lead you to Antares in Scorpius. In my sketch it looks like a bent line, but the dome of the sky is curved, so straight lines are subjective. The constellation of Cygnus the swan is fairly easy to find, looking rather like a cross in the sky. Going back to the Big Dipper, if you follow the curve of its "handle" and continue it across the sky it leads to Arcturus in Boötes, and then on to Spica in Virgo. The phrase to remember is "Arc to Arcturus and spike on to Spica".

Deneb is a blue-white supergiant about 2,700 light-years distant from us; Vega is 25 light-years away and was the pole star 14,000 years ago and is about twice our Sun's size; Altair is a main-sequence star less than 17 light-years away;

Antares is about 555 light-years away and is a red supergiant with a small blue-white main-sequence companion star; Arcturus is an orange giant that is 36.7 light-years away; Spica is a binary system 261 light-years away, whose stars orbit each other in 4 days and they are so close to one another that they pull each other into egg shapes.

Centaurus: This constellation is normally believed to represent an uncivilized centaur (unlike gallant Chiron who is represented by Sagittarius).

Lupus: This was originally seen as an animal being sacrificed by the centaur (Centaurus) but it later came to represent a wolf.

Ara: At the time of the war with the Titans, Zeus made sacrifices on an altar made by the Cyclopes, which were giant one-eyed creatures, and Zeus placed the altar in the sky as the constellation Ara, where it has the Milky Way looking like smoke rising from it.

Sagittarius: Chiron the centaur was accidentally shot by one of Hercules's poisonous arrows and he ended up in unbearable pain. Chiron then gave up his immortality and offered himself as a substitute for Prometheus who was being punished for giving fire to mankind, and Zeus/Jupiter placed Chiron in the sky in the form of Sagittarius.

Scorpius: This is the scorpion that battled Orion and the two got placed in the sky out of sight of each other. (See the Winter Constellations for more of the story).

Virgo and Libra: Virgo the maiden was named Astraea and was a daimone, which were invisible spirits that guarded mortal men during the Golden Age. Astraea is believed to have been the daimone of justice, and Libra was her scales of justice.

Ophiuchus: Chiron (the centaur represented by Sagittarius) was a healer and he taught Asclepius (Ophiuchus) to be a physician. Asclepius/Ophiuchus became such a good healer that one of his remedies would even bring the recently dead back to life but that annoyed Hades, the god of the dead. Hades got Zeus to stop him, and Zeus killed him with a thunderbolt, but then relented, restoring him to life as an immortal and placing him in the sky.

Hercules: Hercules (or Heracles) was Zeus's son, of course, and his mother was the mortal Alcmene who was herself the granddaughter of Perseus and Andromeda. Zeus's wife, Hera, was jealous and tried to kill Hercules several times, but failed. When Hercules married Megara, Princess of Thebes, and had three sons, Hera turned him insane and caused him to kill his wife and children. As punishment for the killings, he was given twelve impossible tasks or labors to complete, during which he ended up holding the heavens on his back for a time, while Atlas was helping him out.

Lyra: Orpheus played the lyre so beautifully that it even tamed wild animals. The lyre had been invented by his father, Apollo. When Orpheus's wife, Eurydice, died after being bitten by a snake, Orpheus used his music to charm his way into the land of the dead (even charming the three-headed dog Cerberus). Orpheus was allowed by Persephone to lead Eurydice out provided that he trusted that she was following and didn't look back, but he couldn't resist making sure she was still there, and she vanished in a puff of smoke. Zeus placed the lyre in the sky as a tribute after Orpheus died.

Corona Borealis: This is the crown of princess Ariadne of Crete. Ariadne had helped Theseus find his way out of the maze where he killed the Minotaur; she had given Theseus a big ball of string that he unraveled while walking through the maze, and then he followed it to find his way out. Sometimes simple technology is the best. Her crown was placed in the sky after she died.

Boötes: Icarius worked out how to cultivate grapes and make wine. As a reward, he was placed in the sky as the constellation Boötes after he died. Sounds like a very worthy task to be immortalized for.

Autumn Constellations

In the autumn we still have the Summer Triangle in the sky, and if you continue the line from Vega to Deneb across to the East, going about twice as far as the distance between Vega and Deneb, you might be able to see a smudgy little spot that is M31, also known as the Andromeda Galaxy. It is called the Andromeda Galaxy because it lies on the northern side of the Andromeda constellation, which itself is attached to the constellation Pegasus. The Great Square of Pegasus stands out quite well in the autumn sky.

Capricornus: When the giant Typhoeus or Typhon suddenly appeared, Bacchus jumped into the Nile and transformed the part of himself that was below water to be like a fish, so he could swim, and the part of him above water into a goat to battle Typhoeus who was trying to tear Zeus/Jupiter to pieces. Zeus then placed Bacchus in the heavens in the form of Capricornus.

Aquarius: Ganymede, the son of the king of Troy, was tending his father's flocks when he was seen by Zeus/Jupiter. Zeus transformed himself into a large bird, flew down, and carried Ganymede to the heavens where he serves as the god's cupbearer, and is also known as Aquarius the watercarrier.

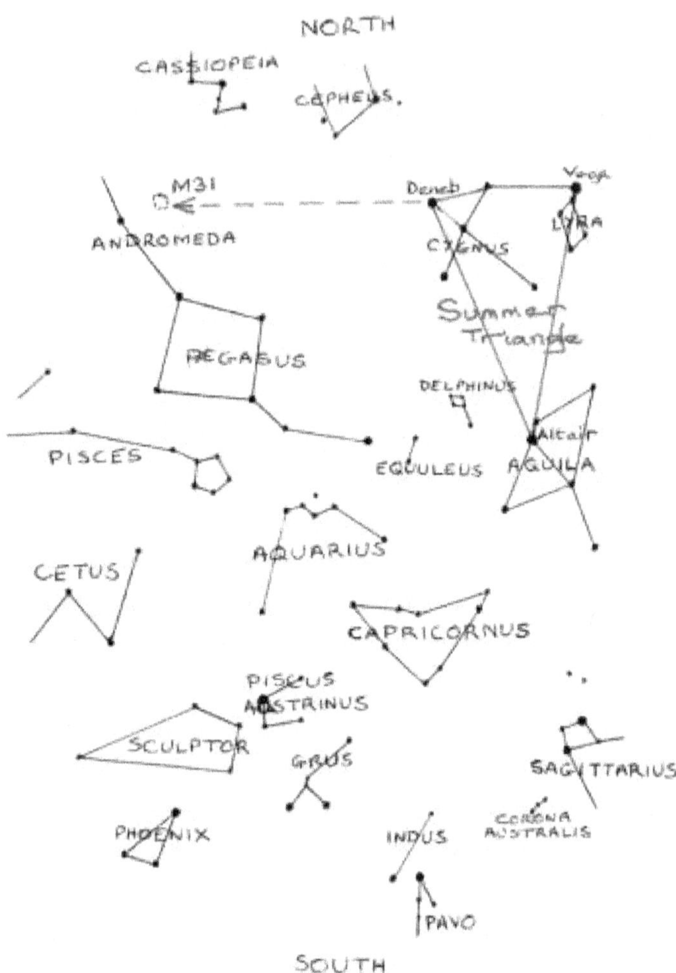

Pisces: When Typhon/Typhoeus suddenly appeared and started attacking the gods, Aphrodite/Venus and her son Eros/Cupid had been on the banks of the river Euphrates, and they turned themselves into fishes to escape. Minerva memorialized the event by placing the image of two fishes in the sky.

Pegasus: Pegasus was a winged horse, and son of the Gorgon named Medusa. Bellerophon tamed it and tried flying on it to heaven, but Zeus sent a horsefly which caused the horse to buck and throw Bellerophon back down to Earth while the horse remained in the heavens.

Delphinus: Poseidon sent a dolphin to convince Amphirite (a Nereid) to marry him, and when she agreed, the dolphin was rewarded by being placed in the heavens as Delphinus.

Sagitta and Aquila: This is related to the story of Chiron, represented by Sagittarius; the punishment that Prometheus was going through was to be pecked at continually by an eagle (sometimes called a vulture), but Hercules shot the eagle or vulture with an arrow, and the arrow was placed in the sky as Sagitta (a small constellation a bit above Altair in the image) and the eagle as Aquila.

Cygnus: One of the stories says that this constellation represents the swan that was Queen Cassiopeia's pet.

Winter Constellations

In the winter sky we have Castor and Pollux curving around to Procyon and on to Sirius (the brightest star in the night sky). We also have the constellation Orion in the sky with its well-known three stars that make up Orion's Belt. If you use those three stars as a pointer in the southeast direction you get to Sirius again, or in the northeastern direction it will take you to Aldebaran in the constellation Taurus. The brightest star on the southern end of Orion is called Rigel, and the one on the northern edge is called Betelgeuse (which is the largest star visible with the naked eye). If you imagine a line from Sirius through Betelgeuse it will curve around towards Capella in Auriga.

These same stars also give us the Winter Hexagon, where you can start with Rigel in Orion, head to Sirius in Canis Major, on to Procyon in Canis Minor, then we get a two-for-one deal as we arrive at Pollux and Caster in Gemini, continue on to Capella in Auriga, then down to Aldebaran in Taurus, and then head back to Rigel.

Aldebaran is an orange giant, 65.2 light-years distant; Rigel is a triple-star system, 864.3 light-years away, with its primary star being a blue-white supergiant; Betelgeuse is a red supergiant, 642.5 light-years distant and it is getting close to collapsing at which time it will explode as a supernova; Capella is actually four stars in two binary pairs and is 43 light-years away.

Orion: After Irieus (a shepherd) offered hospitality to Zeus/Jupiter and Poseidon/Neptune (without realizing they were gods) he was given Orion as a gift. Orion grew up to be a great hunter, and fell in love with Artemis/Diana, the moon goddess who guided the moon across the sky. Her brother, Apollo, complained to Gaia about Orion and Artemis killing too many animals, so Gaia sent the scorpion (Scorpius) after them, and Orion and Scorpius both ended up dying in the ensuing battle. Zeus placed both of them in the heavens, but not where they could see each other.

Canis Major & Canis Minor: The hunting dogs belonging to Orion the hunter, and they were placed in the sky with Orion after he was killed by the scorpion.

Lepus: A hare or jack rabbit that Orion and his dogs chase across the sky (it's less trouble than a scorpion).

Cetus: Sometimes seen as a whale, or as the sea monster that Perseus killed to save Andromeda.

Auriga: Erichthonius, who had the lower body of a snake, was born to Hephaestus and Mother Earth, and he eventually became king of Athens after Athena raised him as her own son. He also invented the four-horse chariot and, as a consequence, after his death Zeus and/or Athena placed him in the heavens as the constellation Auriga, the charioteer.

Taurus: Zeus/Jupiter became enamored with Europa (princess of Phoenicia) and took on the form of a beautiful bull which caught Europa's eye and they played together on the beach. Then Europa tried riding him, and he kidnapped her by swimming out to sea with her, taking her to Crete where he revealed his true self. Taurus represents that bull.

Aries: The cloud nymph Nephele sent a golden-fleece ram to rescue her two children, Phrixus and Helle, because their father (Athamas, king of Boeotia) was trying to have them killed. They flew on the ram's back, but Helle fell off and died in the sea. Phrixus hung on until reaching safety at Colchis, where the ram shed its coat and flew up to the sky and became the constellation Aries. Its fleece was the golden one that Jason and the Argonauts went searching for.

Eridanus: Phaethon was like a youth who wanted to drive his dad's car but wasn't ready. Actually, his dad was Helios the Sun, and the car was the chariot of the Sun. Phaethon couldn't control the horses and the chariot ended up flying too high at times and too low at others, alternately freezing or scorching the Earth. To save the Earth, Zeus hurled a thunderbolt that knocked Phaethon out of the chariot, and he fell into the river Eridanus and drowned. The gods placed the river in the sky to memorialize the event.

Southern Circumpolar Constellations

The southern constellations are relatively recent additions, as you can tell from names like Microscopium and Telescopium. At least, as far as we know the ancient Greeks didn't have microscopes and telescopes, although, since they were able to come up with the Antikythera Mechanism, it seems that they had the techniques and abilities for building very sophisticated pieces of equipment. Anyway, the southern sky does have some interesting constellations, such as Crux (the Southern Cross) even if they are on the small side and lack much in the way of mythology to go along with them. To compensate for that, there are also the two Magellanic Clouds that are irregular dwarf galaxies orbiting the Milky Way galaxy.

The Moon

The Moon is also something you can see most nights, and also during the day at times. You can follow it during the month as it appears first as the thin crescent of a new moon, waxes and grows larger and becomes a full moon a fortnight later (watch out for werewolves if you are out that night), and then wanes and thins out again. It can be interesting to learn the names of the Moon's seas or maria (which are actually large, dark, basaltic plains) and some of the larger craters.

The Moon's main features:

1. Tycho crater
2. Mare Imbrum (Sea of Showers)
3. Mare Serenitatis (Sea of Serenity)
4. Mare Tranquillitatis (Sea of Tranquility) – where man first landed on the Moon (Apollo 11)
5. Mare Fecunditatis (Sea of Fertility)
6. Mare Crisium (Sea of Crises)
7. Oceanus Procellarum (Ocean of Storms)
8. Copernicus crater

There are two orbital periods that we'll be referring to: Sidereal and Synodic. The Sidereal Period is the time to complete one orbit relative to the "fixed" background stars from the point of view of someone on the planet or moon, while the Synodic Period is the time to complete one full pattern of sky motion and get back into the same position in the sky relative to the Sun, as seen from Earth. To remember the difference between them, notice the "real" in "Sidereal" even if it's not pronounced the same way; that period is the real time it takes to complete an orbit. The other one is what the orbital period looks like to us as we are also moving in our orbit around the Sun.

The Moon's real orbital period (its Sidereal period of 27.3 days) differs by a couple of days from the lunar month (new moon to new moon, known as its Synodic period of 29.53 days) because the Earth is also moving in its orbit and the Moon has to go a bit further than one actual orbit to get back between us and the Sun. It orbits Earth in gravitational "tidal lock" keeping the same side turned to Earth, which means that its rotational period is equal to its sidereal period.

A New Moon is when the Moon is between Earth and the Sun, or we could say it is in conjunction with the Sun. The Waxing Crescent can be seen in the west after sunset and appearing higher in the sky as the days pass. The First Quarter Moon shows the "Western half" (from our viewpoint), and it sets near local standard midnight. The Waxing Gibbous Moon sets in the early hours of the morning. The Full Moon occurs when the Moon is in opposition to the Sun, and it is visible all night, rising as the Sun is setting and setting as the Sun is rising. The Waning Gibbous Moon rises in the hours before midnight. The Last Quarter Moon has the "Eastern" half illuminated and rises near local standard midnight. The Waning Crescent Moon can be seen in the east before sunrise.

When the Moon passes through the Earth's shadow, we get a lunar eclipse. At those times the Moon is not likely to go black and disappear from view, but might take on a reddish hue, lit by red light that gets bent as it passes through Earth's atmosphere. Then we get a Red or Blood Moon, but there is also something called a Blue Moon. That can be a bit disappointing, because the Moon doesn't turn blue. A Blue Moon is a second full moon in one month, and that happens fairly rarely (about once a year) so if something is infrequent it is often said to occur "Once in a Blue Moon". Lunar eclipses can be observed from any location at least once a year (clouds permitting), but solar eclipses (when the Moon passes in front of the Sun from our viewpoint), happen a lot less frequently. When a solar eclipse happens, you don't want to normally look directly at it until the main event, or you could blind yourself, but we'll discuss ways to view such events later, when we talk specifically about eclipses.

The terminator is the boundary between the bright and dark regions of the Moon. If it is a waxing Moon, we call it the Sunrise terminator, and with the waning Moon it is called the Sunset terminator. Cusps are where the Terminator meets the edge of the lunar disk (the Lunar Limb), so we get the North and South Cusps. The 3-dimensional nature of features are best observed along, or close to, the Terminator, because shadows become longer there, making the features stand out more. The Full Moon is best for observing ejecta rays emanating from some large craters.

While one side of the Moon always faces Earth, that doesn't mean that you can only see 50% of the Moon. The Moon appears to wobble a bit and, as a result, about 59% of the lunar surface is observable from Earth over time. The wobble is largely related to the fact that the Moon's orbit is an ellipse, so it moves faster or slower than average at times. That results in its rotation seeming to advance occasionally, or fall back at others, letting a bit more of its surface be seen. There are complex east-west and north-south wobble cycles that let you see more than just the normal face, and if you try Googling "moon libration", you will hopefully find the best times to see those usually hidden parts.

Planets

You might have difficulty knowing where to look in the sky for a particular planet, but the planets and the Moon all follow roughly the same path through the sky, and if you know when the Moon is close to a planet it is easier to locate that planet. If you are interested in finding and viewing the planets, go to www.EarthSky.org, and you can search for their monthly guide to where the planets, etc., are, and it gives a lot of information about what you can see that month. Astronomy magazine also has a section each month telling you what you can find where in the night sky.

There are five planets that are visible with the naked eye, namely Mercury, Venus, Mars, Jupiter, and Saturn. Venus is especially visible because (1) it can be the closest planet to us, (2) it is covered in highly reflective clouds, and (3) it is fairly close to the Sun, so it receives a lot of sunlight to reflect on to us. The problem is that Venus, like Mercury, is inside our orbit so, when we look at it, it is always fairly close to the Sun and only appears in the night sky just after sunset or just before sunrise, never in the middle of the night. When either Venus or Mercury get too close to the Sun as viewed from Earth, you won't be able to see them, but that's the same with all planets – it just happens more frequently with Mercury and Venus.

Mercury and Venus are sometimes called the Inferior Planets. The name "inferior" is no slight on them but does indicate that they have distinctive sky behavior. They will always appear near the Sun, alternating between appearing east and west of the Sun, and Mercury is the hardest to find because it is closest to our star, and it is nowhere near as bright as Venus.

If you have a telescope or powerful binoculars you can actually see phases on Venus, like the phases of the Moon. Mercury shows phases too, although they are more difficult to spot because Mercury is smaller and further away. Mercury and Venus will appear in the evening sky after sunset when they are east of the Sun and appear in the morning sky before sunrise when they are west of the Sun.

The elongation of our "inferior planets" is the angle between the planet and the Sun, as seen from Earth. "Greatest Elongation" (east or west) occurs when the line of sight from Earth to the planet forms a tangent to the planet's orbit, at which time the planet is at quarter phase (half illuminated). With Venus, the greatest elongation is always near to 46°, and with Mercury it varies from 18° to 28° due to that planet's highly eccentric orbit. It is easiest to see these planets when they are at or near greatest elongation because they will appear higher in the night sky.

With the inferior planets, planetary disk size varies noticeably with the planet's phase. They will be smallest at Superior Conjunction (Full phase, when they are furthest from us, basically on the far side of the Sun), and largest at Inferior Conjunction (New phase, when they are closest to us). Venus stays near magnitude -4 when visible to us because its apparent disk size compensates for phase and distance — as it gets smaller in size it is also reflecting more light towards us, compensating for the size change. Remember that a star or planet has larger numbers for its "magnitude" as it gets dimmer, and the dimmest star you are likely to see on a really dark night has a magnitude of 6. So, a magnitude of -4 means that Venus is very bright and stands out well even in early twilight. Mercury is brightest in the large gibbous phase and fades considerably in crescent phase, so it is actually brighter from our point of view when it is further away from us.

Inferior planets go through the following synodic cycle: Superior conjunction is when the planet is on the far side of the Sun from Earth and the planet is in its Full phase. At that point we can't see it because it is too close to the Sun in the sky. Then it moves east of the Sun, appearing in the evening west and enters the Gibbous phase. Greatest elongation east is its maximum eastward angle from the Sun, at which point it is in Quarter phase. It then moves back towards the Sun from our viewpoint, sinking in the evening west in a waning Crescent phase. At inferior conjunction it is between Earth and the Sun, and it is in its New phase, so we lose sight of it again. Then it moves west of the Sun, rising in the morning east in a Crescent phase. Greatest elongation west is its maximum westward angle from the Sun, giving us another Quarter phase. It then moves back towards the Sun, appearing in the morning east in a waxing Gibbous phase before getting back to superior conjunction again. When such a planet is in the evening sky, it is catching up with us in its orbit and, when you sight it in the morning, it is racing ahead of us.

Venus spends about 7 months as the "Evening Star," then 2 months out of sight, then about 7 months as the "Morning Star" before disappearing out of sight again for another two months. The best evenings to see Venus or

Mercury are between February and April, and the best morning sightings for these planets occur between August and October. Those months are best because the Ecliptic becomes more vertical to the horizon at those times of year and day, allowing the planets to appear higher above the horizon.

"Superior planets" are those that orbit the Sun further out than Earth, and therefore include Mars, Jupiter, Saturn, Uranus and Neptune. They can appear overhead at midnight, which the "inferior planets" can't. These "superior planets" are normally seen as moving eastward from night to night against the background stars, and that is referred to as "direct motion". Occasionally they appear to move westward against the background stars, which is known as "retrograde motion". That is a visual effect that occurs as we move past them in our smaller and faster orbit. At the beginning and end of this so-called retrograde motion, the planet is said to be "stationary".

Superior planets are always at or very near full phase, from our vantage point. Sometimes their orbits will take them to the far side of the Sun from us, and at that time they are in the sky during our daytime and our view of them is overpowered by the sunlight. Jupiter is a long way away from us, on the other side of the asteroid belt, but it is so big that it still reflects a lot of light back to us and consequently it is the third brightest object in the night sky (after the Moon and Venus).

The Synodic cycle of superior planets (in other words, the way that we see them) is as follows: At conjunction they are on the far side of the Sun from Earth, and they do not appear in the night sky. Then we start to see them rise in the predawn sky, and they will be rising a bit earlier every consecutive night. Later, they appear stationary almost directly overhead against the background stars before beginning their retrograde motion. When they are at opposition they are exactly opposite the Sun, as seen from Earth, or you can say Earth is between the planet and the Sun, and consequently the planet is visible all night. Then the planet will appear stationary again, as it ends its apparent retrograde motion. After that, it begins setting earlier each night, eventually disappearing into the sunset glare, and finally we are back to conjunction.

If a superior planet's Sidereal period (its real orbital period) is less than two Earth years, then its Synodic period (the time for it to get back to the same point in the sky from our perspective) is longer than its Sidereal period, and that is true for Mars which has a Sidereal Period of 687 days or 1.88 years, and a Synodic Period of 780 days or 2.13 years. That results in a long time between oppositions of Mars, which are the best times to send a spacecraft to the planet. For the rest, the synodic periods are much shorter than the sidereal periods. Jupiter's Sidereal period of 11.9 years means that it spends about 1 year in each Zodiacal House, and Saturn's Sidereal period of 29.5 years means that it spends around 2.5 years in each.

If you have a telescope or good binoculars, you can see the four well known Galilean moons of Jupiter. When you have really clear and dark skies, free of artificial lights, technically speaking you should be able to see Jupiter's largest moons with the naked eye, but the light from Jupiter would spoil your eyes' dark vision sufficiently to stop you spotting them. With Saturn, you will definitely need a fairly good telescope in order to see its rings, let alone its moons.

There are six Saturnian satellites that are observable with a reasonable-size home telescope. In order of increasing distance from Saturn, they are Enceladus, Tethys, Dione, Rhea, Titan, and Iapetus. Iapetus has an orbit far beyond Titan's and it is tilted quite a bit from the ring plane, making it the most challenging to observe.

Saturn's bright rings are made of ice and rock. All four of the giant planets have rings, but only Saturn's are viewable in amateur telescopes. The Cassini Division is a gap in the rings that is sometimes viewable in amateur telescopes.

With a telescope, you may be able to see some Jovian or Saturnian satellite events. Transits are when a satellite or moon passes in front of the planet's disk, and Shadow Transits are when the moon's shadow crosses the planet's surface. Eclipses of one moon by another can be observed when one satellite's shadow falls on another, while occultations are when a moon passes behind the planet's disk, disappearing and/or reappearing. You can also get occultations as a result of one moon passing in front of another, giving partial, total or annular occultations. Eclipses of a moon by the planet happen when a moon passes through the planet's shadow, making the moon disappear and/or reappear. Again, you might want to check EarthSky.org's website or subscribe to Astronomy magazine to find out when these kinds of events will be viewable.

The other two giant planets, Uranus and Neptune, are not visible to us without a telescope, although Uranus might just be visible to the naked eye at times but is very hard to spot. You will really need a telescope and know where to look in order to find these two. Forget about finding our dwarf planet, Pluto, unless you have an exceptionally good telescope. And I do mean "exceptionally" good. Even the Hubble Space Telescope could only image it as a few pixels.

Conjunctions of planets and other heavenly bodies have historically been seen as special events, although they happen fairly regularly. The planets and the Moon all travel in roughly the same path across the sky, known as the ecliptic, and they take different times to orbit the Sun, with the consequence that, every so often and not that infrequently, two planets will come close to each other, or the Moon will pass close to a planet, or a planet will come close to a prominent star. Astrologers read special meanings into these events, and some people believe it was a planetary conjunction (or series of them) that became known as the Star of Bethlehem.

Eclipses

Ancient astronomers used eclipses to prove that the Earth was round, from observing the shadow of the Earth during lunar eclipses. Eclipses, both solar and lunar, have enthralled humans for thousands of years, but especially total solar eclipses. Few who see one ever forget it; it can be a tremendously moving experience. I moved all the way up to near Salem, Oregon, for one day to see the total solar eclipse of 2017 that crossed the USA, and it was definitely well worth the trip. It was a truly memorable sight. As daylight dims not that many hours after sunrise and streetlights start to come back on, the Sun has a larger and larger bite taken out of it, until it is all gone, except that these amazing glowing forms appearing in a ring around the black disk where the Sun had been. That ring is the corona, the Sun's violent outer atmosphere, which suddenly becomes visible because it is no longer drowned out by the brightness of the Sun itself.

Having the Sun vanish during daytime, naturally would have been viewed with fear and wonder by those not expecting it and not understanding what was happening. The Chinese would bang pots to scare away the dragon that was eating the Sun, and it worked! The Sun came back. The meanings that people put on the event can vary, but it is easy for fear to break loose when day suddenly and unexpectedly becomes night. Norse mythology talks of Ragnarök, with the end of all things occurring in the cold and dark, so when the Sun started to vanish as Vikings were sailing to the Shetland Islands, they thought the end of the world had come. A battle between the Medes and the Lydians was stopped by a solar eclipse, which was seen as the gods giving a sign, and they made a truce. That eclipse gave a good result and seemed to have been a good omen in that situation, but usually it was seen as an evil omen.

The Babylonians realized, from the records they had kept, that there was an 18-year cycle involved with eclipses, and the Greeks used that information when creating a kind of mechanical "computer" for predicting eclipses and other astronomical phenomena. That remarkable "computer" is the Antikythera Mechanism that I briefly mentioned earlier. The 18-year cycle that the Greeks noticed is the Saros Cycle that says that eclipses recur every 6582.32 days, which is equal to 18 years and 10.32 or 11.32 days (depending on leap years). These recurring eclipses are "almost" identical, but the extra 0.32 days means that the Earth will have rotated almost 8 hours further (about a third of a full rotation). Consequently, with every third Saros cycle, the eclipse will recur in almost the same time and place. However, the universe, and even the Earth-Moon system, is not a perfectly balanced piece of clockwork that doesn't vary, so differences build up over time with the result that a particular Saros series will normally last only around 1,226 to 1,550 years. That's long enough for me anyway.

Once every 16 to 18 months, you can expect to be able to see a solar eclipse somewhere on Earth. The Moon and the Sun are vastly different in size, with the Sun approximately four hundred times bigger in diameter than the Moon, but it is also approximately 400 times further away, so the two appear almost exactly the same size in our sky. However, the Moon is slowly moving away from us, with the result that it will no longer completely block the Sun in half a billion years, so don't wait too long to see an eclipse! What a coincidence that the Moon and Sun fit so neatly right at the time when intelligent species emerged on Earth. At least, we like to think that we are intelligent.

Where you are when a solar eclipse occurs, changes what you will see. There is a fairly narrow band marked out on the Earth where you can see a particular total solar eclipse, and on either side of that are wider bands from which you could see a partial eclipse, with the Moon taking a bite out of the side of the Sun but, since it never fully covers the Sun, the corona is not seen. The Penumbra is the zone on Earth from which you see a partial eclipse, and the Umbra (less than 100 miles wide) is where you see a total eclipse. Any specific place on Earth can expect to see a total solar eclipse about once every 360 to 375 years, and with that time gap you can probably understand how they get to be taken seriously when they occur.

An Annular Solar Eclipse does not occur annually but happens when the Moon is further from Earth than usual in its elliptical orbit, so the apparent lunar disk is smaller than the solar disk. Effectively, the Moon's umbra ends above the Earth's surface, so the viewer is said to be in the antumbra. "Annulus" means "ring" and the viewer sees a bright solar ring around the black outline of the Moon. Because that ring is the edge of the Sun, your eyes cannot adjust to see the much dimmer solar corona during an annular eclipse, and you need to be using eclipse glasses or other very dark filters anyway.

In a solar eclipse, the Moon casts its shadow on Earth so, technically, we see a lunar occultation of the Sun. Officially, an eclipse involves the object that is being viewed from casting its shadow onto another body. The source of light for eclipses and occultations in our solar system is always the Sun, and a viewer on the eclipsed object sees an occultation or transit of the Sun. But semantics aside, solar eclipses are great things to see, even if they are technically occultations.

A lunar eclipse occurs when the Moon passes through the Earth's shadow. So, a Lunar Eclipse is a real eclipse because we see the Earth's shadow darkening the Moon, while (as mentioned above) a solar eclipse is really a lunar occultation of the Sun because that is the Moon blocking the Sun from our view by casting its shadow on us. Lunar eclipses supposedly foretold famine and disease to the Chinese, and bigger or more frequent earthquakes to the Japanese. There was a lunar eclipse on the winter solstice in 2010, and a 7.4-magnitude earthquake occurred in Japan at the same time, but there is no correlation between eclipses and earthquakes seen in long term statistics. Coincidences do occur, about as regularly as statisticians predict but more frequently than most people expect.

Columbus knew eclipses were predictable, and he used that knowledge to impress the natives of Jamaica, forecasting a total lunar eclipse that he already knew would occur and he used that event to get the natives to provide the food and services that he and his crew needed. Alexander the Great used the observation of a total eclipse to encourage his army in Mesopotamia, turning it into a weapon of fear against the Persians, and he won the battle. So, it was a good omen for Alexander, but an ominous one for the Persians.

There are various types of lunar eclipses. A penumbral lunar eclipse is when the Moon passes through Earth's penumbra, but this type of eclipse is usually too subtle to notice, although the darkening can be noticeable during a "deep" penumbral eclipse. A partial lunar eclipse occurs when only part of the Moon's disk passes through Earth's umbra. In a total lunar eclipse, all of the Moon's disk passes through Earth's umbra. One thing to look for is the "Japanese Lantern effect", a special illusion that can occur a few minutes before and after totality and lasting for 30 seconds or less, resulting from the varying brightness from one side of the Moon to the other. The redness of the Moon during a total lunar eclipse depends upon Earth's atmosphere, and the more clouds and/or dust in Earth's terminator region, the redder and darker the Moon will appear.

Lunar eclipses can only occur when the Moon is full, and solar eclipses only at a New Moon. The Moon's orbit is inclined to the ecliptic about 5 degrees, so eclipses can only occur when the Moon's orbit crosses the ecliptic, which is defined as the path that the Sun takes across the sky[cccxliv]. Consequently, there are two 36-day eclipse seasons each year, about 6 months apart, and they occur about 19 days earlier each year.

Meteors

Meteors are normally about the size of grains of sand, but they still make an impressive flash across the sky. Most are believed to come from the asteroid belt and are small remnants resulting from occasional collisions between asteroids, or they may be primordial bits of matter from the days of the early solar system that had never grown any bigger. Some meteors are largely iron, indicating they probably came from the core of a large asteroid or protoplanet that got shattered in a collision. Some meteors have been established as having come from the Moon and others from Mars; these would be the result of a major impact on the Moon or Mars that blasted rocks out into space.

However, the meteors that you can predict, and almost guarantee seeing, are remnants of comets that have passed our way throughout history. These remnants travel in known orbits that follow the path of the original comet, so we know when they are likely to enter our atmosphere. This is how we get the regular meteor showers, such as the Taurids, Perseids, and Orionids. The Geminids meteor shower is the odd-man-out because this one comes from an asteroid named Phaethon which is believed to have its sodium vaporized in the heat as it approaches the Sun, blowing off rock and dust near the surface and spreading the material through space to give us such a nice display in December[cccxlv]. As you might guess, meteor showers are named after the constellation that they seem to emerge from, before they streak across the sky in all directions. Halley's comet is a regular visitor to our region of space, so it is not surprising that two of these meteor showers, the Orionids and the Eta Aquarids, are remnants from that comet. If a meteor (sometimes called a falling or shooting star) is large enough so that part of it manages to land on Earth, the remains become known as a meteorite. Before they enter our atmosphere and burn up, they are known as meteoroids. Fireballs are particularly large meteors, and they are the source of many UFO reports.

Comets

Comets are icy bodies that start to boil off, or outgas, as they get close to the Sun. The tail of a comet is this outgassing being blown by the solar wind. Well, some of the matter is in the form of gas that gets ionized, and that is driven directly away from the Sun by the solar wind as a plasma tail. Meanwhile, the dust-sized particles are affected by the pressure of the sunlight itself and the movement of the comet in its orbit, leading to the brighter and often slightly curved dust tail. It is common to think of the tail as flowing behind the body of the comet as it hurtles through space but, after the comet has swung around the Sun and is heading back out into deep space, the tail will actually be seen blowing ahead of the comet body. It is also sometimes thought that comets flash across the sky similar to meteors, but they can be in the sky for weeks, moving in much the same way as a planet.

Comets can be short-term, and this type comes mainly from the Kuiper Belt out near where Pluto orbits, or the comets can be long-term and are from the Oort Cloud that is calculated to orbit the Sun almost halfway to the nearest star (or the second nearest star, Proxima Centauri – the nearest star is the Sun). Long-term comets have orbits of thousands of years. The short-term ones have orbits that only go up to hundreds of years. Some, like Halley's comet, have had their orbits affected as a result of passing by the planets, especially Jupiter.

Comets come our way fairly regularly, but very few

come close enough for people to see the tail without a telescope or powerful binoculars. Such a comet may only appear in the sky once in a person's lifetime. Sometimes they are even bright enough to be seen in broad daylight, and they can remain visible for weeks or months. When something that rare and unusual occurs, it is understandable that people can come to see it as an omen, and why not make it an *evil* omen, just to be prepared? If a comet got too close to us, it could do as much damage as the asteroid that wiped out the dinosaurs, so they definitely are a potential threat. But they also have been useful to our planet, because they are believed to have been the source of a lot of the water that we have in our rivers and oceans today and many of the organic chemicals that life uses. The most famous comet is Halley's, and that comet appeared at the time of the Norman invasion of England at the Battle of Hastings and is depicted in the Bayeux Tapestry. That would have been a bad omen for one side and a good omen for the other. Halley's Comet will not be back in our sky until mid-2061, but if you get up early one morning to watch the Orionid meteor shower you can say you've seen Halley's Comet (or at least part of it) and you can get to wish on a falling star at the same time.

Zodiacal Lights and the Gegenschein

The plane of the inner solar system contains fine dust and debris, which sometimes enter the atmosphere and that we see as meteors, but on rarer occasions we can see a glow in the sky caused by sunlight reflecting off of this material.

The Zodiacal Lights are where sunlight is scattered off such debris. It needs a really dark sky and might be seen in the west just after evening twilight on moonless nights around February and March, and in the east just before morning twilight on moonless nights around September and October. It is said to look like a bright, tall, tapering triangle of light along the Ecliptic (hence the word "zodiacal" in the name), and it is similar in brightness to the Milky Way. To have a chance of seeing it, you need ideal viewing conditions with a dark night and virtually no light pollution from streetlights, etc., which accounts for why I have never seen it.

The dust creating the zodiacal lights was thought to have been mostly shed by comets and asteroids, but as the Juno spacecraft headed for Jupiter it was bombarded by interplanetary dust particles at around 10,000 mph (16,000 km/hr.), with the cloud of dust stretching from Earth's orbit to just beyond Mars, and the majority of the dust being around Mars's orbit. So, Mars and/or collisions in the asteroid belt may have been its origin, with Earth's gravitational influence sculpting the inner edge, and Jupiter the outer[cccxlvi].

The Gegenschein is where sunlight is reflected back from debris directly opposite the Sun in the night sky. This effect can also be quite difficult to detect, and it requires dark moonless nights and very clear skies with no light pollution around. It is said to be best seen around midnight during either late September to early November, at which time you could find it in Pisces, or late January to early

February, when it will be in Cancer. Look for an elliptical glow centered on the Ecliptic.

Auroras

We have probably saved the best for last.

The Aurora Borealis or Northern Lights, and the Aurora Australis or Southern Lights for our antipodean readers, are caused by energetic and polarized particles of plasma from a solar flare being caught by Earth's magnetic field and directed towards the polar regions, where the magnetic field dips down through our atmosphere. As the energetic particles interact with the gasses in the atmosphere, they can cause a brilliant light display that is reminiscent of pulsing neon lights, giving a spectacular show. To the ancient Romans, Aurora was the goddess of the dawn, but was not specifically associated with the Northern Lights. To the Australian Aborigines, the Aurora Australis is associated with fire, and sometimes was seen as an evil spirit creating fire. Some Native Americans saw auroras as the spirits of their departed friends dancing in the sky, which could be a good sign. They are seen as bad omens by the Inuit, and solar astronomers watch out for them because, if the effect is too strong, it can disrupt the electrical grids and knock out the satellites that we rely on for communication and navigation.

Green is a very common color for auroras, and that comes from interaction with oxygen atoms at lower levels, while at higher altitudes oxygen will produce red auroras. Nitrogen normally produces blue, and sometimes red.

Other gasses can add to the colors.

Appendix B – Mysteries

You probably noticed that we are still uncertain about some very important issues in the life of the universe, and I'd like to list some of the big ones:

- We don't know what caused the Big Bang or if it started with a singularity, or if singularities can exist.
- We don't understand gravity, whether there are gravitons, and why gravity is so weak compared to other forces.
- We don't know if inflation happened and, if it did, how it did, and we don't know whether it was really needed, how it might have started, or how it stopped in our universe.
- We don't know why matter and antimatter didn't annihilate each other completely, in other words, we don't know why there's any matter.
- We don't know what role magnetic fields played in shaping the universe.
- We don't know if the universe is finite or infinite.
- We know almost nothing about the nature of Dark Matter and Dark Energy except for their effects.
- We don't know what constitutes "the fabric of space".

- We don't know why the cosmological constants are the values that they (thankfully) are.
- We don't know if the universe is really "flat".
- We don't know if Hawking radiation is real.
- We don't know how life started on Earth or if it exists elsewhere in any form.
- We don't know if life originated here on Earth, or if it might have come from elsewhere.
- We don't know when the first "animals" – sponges and/or comb jellies – appeared on Earth.
- We don't know if negative strange matter has already started transforming the universe, or if it can actually do that.
- We don't have a good way of selecting which is the most likely model for how the universe will end, even assuming that any of our current ideas are right.

The JWST and the upgraded LHC may answer some of these questions and maybe render a lot of what I've described as "facts" to be fiction. What I've written about may work in models and might even have been used to create a universe or two, just maybe not ours. What we can be certain of is that new questions will certainly be raised as we come up with new answers to the old questions. The developing field of gravitational wave astronomy is also expected to continue turning up surprises.

Yet, despite all this missing knowledge, we can still be proud of how far we have come in piecing together the history of the universe. We've certainly learned enough to know of a lot of things and identify things that we don't know. The big question is: What haven't we identified yet and know absolutely nothing about? These are exciting times, and we have a lot more to get excited about in the years ahead. It's in the stars!

Astronomical Alchemy

About the Author

Geoff was eleven years old when Sputnik 1 launched the Space Age, and fourteen when Yuri Gagarin lifted off in Vostok 1 to start the era of manned space flight. He had been interested in science and astronomy from an early age, and to find himself in the opening era of space exploration and a time of so many developments in the sciences was extremely exciting to him.

Geoff received his MSc in IT in 2005 through the University of Liverpool, and his PhD in metaphysics in 2013 through the Metaphysical Institute. His interests include almost all areas of science, especially neuroscience, cosmology, and quantum physics. Since human spaceflight began when he was still at a very impressionable age, it has been something he has followed closely, which then led on to space and astronomy. With that background, you might have also guessed that he is an avid Star Trek fan (all iterations of the series).

He now lives in the wine country of northern California with his beautiful wife Mary, enjoying the view of the mountains and watching the local deer, jack rabbits and other local wildlife. An idyllic lifestyle, interrupted occasionally by a rapid evacuation as a wildfire comes over the mountains.

You can reach Geoff at gcanham@compuserve.com. He is @GeoffCanham on Twitter, and you can find him on Facebook at www.facebook.com/geoff.canham, although he doesn't make a lot of use of social media. No 5 a.m. Tweets from him.

His first book was "Free Will or Won't … are we in charge of our lives or not?", and his second was "It's About Time … Relatively Speaking". Both are available in paperback and Kindle ebook formats.

His Goodreads page can be found by Googling "Goodreads Geoff Canham" or go directly to:
  www.goodreads.com/author/show/18839749.Geoff_Canham

His Amazon Author page is at www.amazon.com/-/e/B07TFBL3FC

References

The following sources of information probably doesn't list all those that I have accessed while gathering the information used in compiling this book, but I have tried to give credit where I can. You'll notice Astronomy magazine, New Scientist magazine, and the courses from TheGreatCourses.com appearing many times, and Wikipedia, of course, and I can really recommend them all for anyone interested in keeping up with the latest discoveries regarding astronomy and cosmology. You'll also find some fascinating books within the list. And a special shout-out for the Robert Ferguson Observatory – support your local observatory and astronomy club.

[i] Cosmos: A Spacetime Odyssey – Standing Up in the Milky Way, aired 9 March 2019
[ii] Astronomy February 2022, Michael E. Bakich, Meet 20 Great Astronomers
[iii] Wikipedia, Fritz Zwicky, accessed 14 May 2021
[iv] Wikipedia, Nancy Roman, accessed 14 May 2021
[v] Wikipedia, Vera Rubin, accessed 14 May 2021
[vi] Astronomy January 2021, Dan Hooper, It Began With the Big Bang
[vii] The First Three Minutes, Steven Weinberg, 1977, 1993
[viii] New Scientist 2 January 2021, Stuart Clark, Measuring Up The Universe
[ix] Astronomy April 2022, Stan Odenwald, Imagining Our Infant Universe
[x] New Scientist 24 July 2021, Chandra Prescod-Weinstein, The

eternal debate about eternal inflation

[xi] ScienceDaily 25 May 2021, University of Copenhagen – Faculty of Science, New Details on What Happened in the First Microseconds of the Big Bang

[xii] New Scientist 26 June 2021, Ian Taylor, Primordial Magnetism

[xiii] Astronomy, January 2021, Christopher Conselice, The Emergence of Matter

[xiv] Astronomy Magazine Special October 2020 Cosmos – Origin and Fate of the Universe, Dan Hooper, Is The Big Bang in Crisis

[xv] New Scientist 13 August 2022, Leah Crane, Gamma ray bursts could unpack universe expansion

[xvi] Astronomy July 2021, Vitthal Nath, Ask Astro

[xvii] The Great Courses, Prof. Alex Filippenko, Understanding the Universe: An Introduction to Astronomy, 2nd Edition, published 2007

[xviii] New Scientist 14 May 2022, Leah Crane, Distance Unknown

[xix] New Scientist 29 January 2022, Leah Crane, Clashing figures for universe's growth are starting to look more serious

[xx] New Scientist 24 July 2021, Leah Crane, We may finally solve the mystery of how fast the universe is expanding

[xxi] New Scientist 3 July 2021, Terry Tucker, The Back Pages – Almost the Last Word – Expansive Objects

[xxii] Astronomy January 2021, Avi Loeb, Exploring the Shape of Spacetime

[xxiii] Killers of the Cosmos, series 1 episode 6, Big Sleep

[xxiv] Astronomy January 2021, Brian Keating, Inflating the Universe

[xxv] The Great Courses, Prof. Gary Felder, The Big Bang and Beyond: Exploring the Early Universe, published 2022

[xxvi] Our Mathematical Universe, Max Tegmark

[xxvii] New Scientist 1 December 2012, Marcus Chown, In The Beginning

[xxviii] New Scientist 22 September 2018, Joshua Howgego, The Mystery of the Universe in 10 Objects – 1 Cosmic Microwave Background

[xxix] Astronomy June 2021, Arwen Rimmer, The Age of Multi-Messenger Astronomy

[xxx] How the Universe Works, Series 9 episode 8, Secret Lives of Neutrinos, aired 12 May 2021

[xxxi] Astronomy March 2021, Geoffrey Mo, Sizing Up Black Hole Mergers
[xxxii] Astronomy Magazine Special October 2020 Cosmos – Origin and Fate of the Universe, Jesse Emspak, Relativity – Right or Wrong?
[xxxiii] New Scientist 23 January 2021, Leah Crane, A Sea of Gravitational Waves?
[xxxiv] Astronomy May 2021, Mark Zastrow, Pulsars hint at background sea of gravitational wave
[xxxv] New Scientist 4 September 2021, Leah Crane, Waves in spacetime could let us see if gravity is quantum
[xxxvi] New Scientist 5 February 2022, Leah Crane, Gravitational wave echoes could reveal dark matter
[xxxvii] How the Universe Works, Series 9 episode 10, Gravitational Waves Revealed, aired 26 May 2021
[xxxviii] New Scientist 10 October 2021, Paul M. Sutter, Gravity with a twist
[xxxix] New Scientist 15 January 2022, Harry Cliff, Dawn of a new physics?
[xl] New Scientist 19 June 2021, Leah Crane, … while evidence for dark matter mounts in Milky Way
[xli] New Scientist 19 June 2021, Bas den Hond, Dark matter fails key galaxy merger test …
[xlii] New Scientist 6 November 2021, Leah Crane, Physicists fail to find sterile neutrino
[xliii] Wikipedia, Axion, accessed 1 December 2021
[xliv] Astronomy September 2021, Robert Lea, Dark Matter – the unusual suspects
[xlv] QuantaMagazine.org, Markos Kay & Jennifer Ouellete, Dark Matter Recipe Calls for One Part Superfluid, published 13 June 2017
[xlvi] Wikipedia, X17 particle, accessed 1 December 2021
[xlvii] Through the Wormhole, Series 2 episode 4, Are There More Than 3 Dimensions?, aired 29 June 2011
[xlviii] New Scientist 8 June 2019, Michael Brooks, Welcome to the mirror world
[xlix] BBC.com, Michio Kaku, String Theory, accessed 15 August 2022
[l] New Scientist 5 June 2021, Dark matter map is most extensive

yet

[li] Astronomy January 2021, Bruce Dorminey, The Mystery of Dark Energy

[lii] Astronomy April 2021, Andrew Friedman, Ask Astro - Expanding Space

[liii] Astronomy January 2021, Dana Najjar, The Cosmic Dark Ages

[liv] Wikipedia, Recombination (cosmology), accessed 15 November 2021

[lv] New Scientist 22 February 2022, Chandra Prescod-Weinstein, How does fusion work?

[lvi] Astronomy January 2021, Michael E. Bakich, The First Stars Are Born

[lvii] How the Universe Works, Series 8 episode 75, Hunt for Black Holes, aired 29 September 2021

[lviii] Astronomy Magazine Special October 2020 Cosmos – Origin and Fate of the Universe, Mara Johnson-Groh, Dark Stars Come Into The Light

[lix] New Scientist 16 April 2022, Leah Crane, Astronomers have spotted the most distant galaxy ever

[lx] New Scientist 30 July 2022, Jonathan O'Callaghan, JWST spots oldest ever galaxy …

[lxi] Astronomy June 2021, 1.3 Million

[lxii] Astronomy April 2021, Glenn Chaple, Color-coding stars

[lxiii] Astronomy June 2021, Caitlyn Buongiorno, Stellar Tantrum

[lxiv] Astronomy March 2021, Alison Klesman, Mystery of the Blue Ring solved

[lxv] Astronomy May 2020, Jake Parks, Quick Takes, Remind Me Later

[lxvi] New Scientist 25 December 2021, Really brief – Vast eruption seen on a sun-like star

[lxvii] Astronomy April 2021, Steve White, Ask Astro

[lxviii] Astronomy February 2021, Alison Klesman & Jake Parks, Top 10 Space Stories of 2020

[lxix] Astronomy July 2021, Mark Zastrow, Another Star Sneezes Like Betelgeuse

[lxx] How the Universe Works, Series 9 episode 5, Curse of the White Dwarf, aired 21 April 2021

[lxxi] New Scientist 9 October 2021, Jonathan O'Callaghan, Weird

white dwarf that is too cold defies explanation
[lxxii] New Scientist 10 July 2021, Leah Crane, Tiniest dwarf star may go out in a bang
[lxxiii] Astronomy February 2022, Alison Klesman, Some white dwarfs can look young again
[lxxiv] Astronomy April 2020, Isaiah Charnow, Ask Astro
[lxxv] Wikipedia, Variable star, accessed 25 February 2022
[lxxvi] How the Universe Works, Season 5 episode 5, Stars That Kill, aired 10 January 2017
[lxxvii] New Scientist 4 September 2021, Jason Arunn Murugesu, Cosmic rays from an 'ordinary' star explosion
[lxxviii] Astronomy August 2021, Alison Klesman, Titanium debris could solve supernova mystery
[lxxix] Astronomy October 2021, Alison Klesman, Third type of supernova explosion confirmed
[lxxx] New Scientist 18 December 2021, Leah Crane, 'Space cow' explosion was probably a failed supernova
[lxxxi] New Scientist 11 September 2021, Leah Crane, Strange star death spotted in a galaxy far, far away
[lxxxii] Astronomy Magazine Special October 2020 Cosmos – Origin and Fate of the Universe, Liz Kruesi, Supernova 1987A 30 Years Later
[lxxxiii] New Scientist 22 January 2022, Alex Wilkins, Giant bubble around solar system charted
[lxxxiv] Astronomy May 2022, Christopher Cokinos, A 1,000-light-year-wide cosmic bubble surrounds Earth
[lxxxv] Astronomy Magazine Special October 2020 Cosmos – Origin and Fate of the Universe, David J. Eicher, A Universe of Galaxies
[lxxxvi] Astronomy Magazine Special October 2020 Cosmos – Origin and Fate of the Universe, Christopher J. Conselice, Our Trillion-Galaxy Universe
[lxxxvii] New Scientist 23 January 2021, The Back Pages, Almost the Last Word
[lxxxviii] Astronomy May 2020, Jake Parks, Quick Takes - Baker's Dozen
[lxxxix] Astronomy March 2021, Dennis Moore, Ask Astro
[xc] New Scientist 12 June 2021, Leah Crane, "Cosmic collisions may push huge black holes off-kilter"

[xci] Astronomy April 2020, Alison Klesman, One Galaxy, Three Supermassive Black Holes
[xcii] New Scientist 11 December 2021, Leah Crane, Massive black holes heading for collision
[xciii] Astronomy April 2021, Alison Klesman, Astronomers Spot a Galaxy in Midlife Crisis
[xciv] Astronomy April 2021, Caitlyn Buongiorno, How Supermassive Black Holes Kill Their Galaxies
[xcv] Astronomy January 2021, Michael E. Bakich, How to Build a Galaxy
[xcvi] Astronomy May 2020, Jake Parks, Quick Takes, Early Bird Special
[xcvii] Astronomy March 2021, Caitlyn Buongiorno, Mature Galaxies Populated the Early Universe
[xcviii] New Scientist 19 June 2021, Leah Crane, Vast black holes clear way for some galaxies to thrive
[xcix] Astronomy Magazine Special October 2020 Cosmos – Origin and Fate of the Universe, Roen Kelly, Do All Galaxies Have Dark Matter?
[c] Astronomy April 2020, Jake Parks, More Galaxies Found to be Missing Dark Matter
[ci] Astronomy April 2022, Jake Parks, More Missing Dark Matter
[cii] Astronomy May 2021, Hailey Rose McLaughlin, When Worlds Collide
[ciii] Astronomy May 2021, Mark Zastrow & Jake Parks, Quick Takes
[civ] Astronomy April 2020, Erika A. Carlson & Alison Klesman, A Galactic Collision
[cv] Astronomy June 2021, Hailey Rose McLaughlin, People Are Strange, Galaxies Are Stranger
[cvi] Astronomy May 2020, Alison Klasman, Our Nearly Gentle Giant
[cvii] Astronomy June 2021, Jake Parks, Quick Takes – Gravity Gang
[cviii] Astronomy November 2021, Caitlyn Buongiorno, Cluster Sports Abundance of Black Holes
[cix] EarthSky.org, Kelly Kizer Whitt, What are 'open' star clusters?, accessed 15 February 2022
[cx] New Scientist 31 July 2021, Leah Crane, Cosmic collisions could outshine supernovae
[cxi] Astronomy May 2020, Alison Sills, Ask Astro - Stellar Collisions

[cxii] Astronomy August 2021, Alison Klasman, Ask Astro
[cxiii] Astronomy February 2021, Robert Harrison, Ask Astro
[cxiv] Astronomy February 2021, David Martinez-Delgado and R. Jay GaBany, Studying Galaxies with Amateur Images
[cxv] New Scientist 16 April 2022, Leah Crane, Fast-moving stars probably come from other galaxies
[cxvi] New Scientist 25 December 2021, Thomas Lewton, Galactic Ghosts
[cxvii] Astronomy December 2021, Hailey Rose McLaughlin, The Milky Way has a broken arm
[cxviii] New Scientist 21 August 2021, Leah Crane, Huge stream of gas may be unknown arm of our galaxy
[cxix] Astronomy August 2021, Mark Zastrow, Quick Takes – Light Finds a Way
[cxx] Astronomy April 2020, Jake Parks, When the Milky Way erupted in 100,000 Supernovae
[cxxi] New Scientist 5 June 2021, Leah Crane, Features Interview with Andrea Ghez
[cxxii] New Scientist 19 December 2020, Leah Crane, Vast gas balloons may have been created by black hole
[cxxiii] Astronomy February 2021, Alison Klasman & Jake Parks, 10 Top Space Stories of 2020 - The Milky Way does the Wave
[cxxiv] Astronomy September 2021, Jake Parks, Quick Takes – Cosmic Hum
[cxxv] How the Universe Works, Season 5 episode 6, The Universe's Deadliest, aired 17 January 2017
[cxxvi] Astronomy July 2021, Hot Bytes – Turn It Down
[cxxvii] Astronomy February 2021, Alison Klesman and Jake Parks, Earliest Galaxy Group Found
[cxxviii] Astronomy Magazine Special October 2020 Cosmos – Origin and Fate of the Universe, Michael West, Why Do Galaxies Align?
[cxxix] Astronomy Magazine Special October 2020 Cosmos – Origin and Fate of the Universe, Liz Kruesi, Galaxy Clusters – The Universe's Cosmic Lenses
[cxxx] BBC.com, Jonathan Amos, Hubble: 'Single star' detected at record-breaking distance (published online March 30, 2022)
[cxxxi] Astronomy May 2020, Alison Klesman, Quantum Gravity, Shocking Scene

[cxxxii] Wikipedia, The Giant Arc, accessed 4 April 2022

[cxxxiii] Astronomy February 2021, Mark Zastrow, Simulations Zoom in on Dark Matter

[cxxxiv] EarthSky.org, Kelly Kizer Whitt & Deborah Byrd, Astronomers Spot Largest Rotation in the Universe

[cxxxv] Astronomy February 2021, Caitlyn Buongiorno, Tangled in a Cosmic Web

[cxxxvi] Astronomy July 2021, Caitlyn Buongiorno, Strands of the Cosmic Web Revealed

[cxxxvii] New Scientist 30 April 2022, Thomas Lewton, A new slant on the cosmos

[cxxxviii] New Scientist, 25 June 2022, Leah Crane, The universe is weirdly lopsided

[cxxxix] Astronomy May 2020, David J. Eicher & Brian May, Cosmic Clouds

[cxl] Astronomy May 2020, Rob Jeffries, Ask Astro

[cxli] New Scientist 24 July 2021, Sapphire Lally, Where does gold come from?

[cxlii] Astronomy October 2021, Caitlyn Buongiorno, Neutron Stars – A cosmic goldmine

[cxliii] New Scientist 19 February 2022, Jonathan O'Callaghan, Rare 'hyperburst' is powerful nuclear explosion in space

[cxliv] Astronomy April 2020, Alison Klesman, The First Map of a Neutron Star

[cxlv] New Scientist 4 June 2022, Jonathan O'Callaghan, 'Zombie' pulsar is active member of star graveyard

[cxlvi] New Scientist 17 July 2021, Leah Crane, Distant space cloud is firing supercharged photons at us

[cxlvii] Astronomy June 2021, Hot Bytes – Newborn Pulse

[cxlviii] Astronomy May 2021, Caitlyn Buongiorno, Astronomers find the youngest, fastest spinning magnetar yet

[cxlix] Astronomy March 2021, Caitlyn Buongiorno, Magnetic Star Born from a Colossal Collision

[cl] New Scientist 5 February 2022, Leah Crane, Mysterious signal beaming from afar

[cli] Astronomy February 2021, Eric Betz, Source of Fast Radio Bursts Revealed

clii Astronomy May 2020, Alison Klesman, Second repeating fast radio burst tracked, deepening mystery
cliii Wikipedia, Exotic star, accessed 5 September 2021
cliv Astronomy June 2021, Jake Parks, Quick Takes – Boson Buddies
clv New Scientist 1 May 2021, Leah Crane, Antistars may be lurking close by
clvi Astronomy May 2021, Jake Parks, Quick Takes, Black Hole Bullies
clvii Astronomy December 2021, Yvette Cendes, How to Swallow a Star
clviii New Scientist 29 January 2022, Alex Wilkins, Tiny galaxy's black hole drives star birth
clix Astronomy June 2021, Caitlyn Buongiorno, Cygnus X-1 Springs a Massive Surprise
clx New Scientist 22 May 2021, Leah Crane, Searching for the Earliest Black Holes
clxi New Scientist 2 October 2021, James O'Callaghan, Curious craters
clxii Astronomy March 2021, David J. Eicher, The Most Fascinating Monsters
clxiii Astronomy March 2021, Robert Naeye, How to Grow a Giant Black Hole
clxiv Astronomy June 2021, Jake Parks, Quick Takes – Dark Matter Beasts
clxv Astronomy July 2021, Alison Klesman, Event Horizon Telescope Maps a Black Hole's Magnetic Field
clxvi Astronomy April 2020, Jake Parks, Quick Takes, Biggest Black Hole
clxvii Astronomy February 2021, Alison Klesman & Jake Parks, Top 10 Space Stories of 2020, First Midsized Black Hole Detected
clxviii New Scientist 19 March 2022, Stuart Clark, Wave after wave
clxix Astronomy January 2021, Nola Taylor Redd, How Black Holes Die
clxx New Scientist 18 September 2021, Leah Crane, Black holes just got weirder thanks to quantum pressure
clxxi New Scientist 25 September 2021, Paul Davies, The Great Disappearing Trick
clxxii New Scientist 23 April 2022, Leah Crane, Foam and laser help

us probe star birth

clxxiii Astronomy January 2021, Michael E. Bakich, Our Solar System's Origin

clxxiv Space's Deepest Secrets, Series 2 episode 9, Stranger Signals from Alien Worlds, aired 6 June 2017

clxxv New Scientist 13 November 2021, Jonathan O'Callaghan, Did Earth grow from an alien rock?

clxxvi New Scientist 4 December 2021, Stuart Clark, Shaken and stirred

clxxvii New Scientist, 11 December 2021, Will Gates, New look at infant suns may solve galactic mystery

clxxviii Astronomy October 2021, Alison Klesman, Solar System Sizes

clxxix New Scientist 2 April 2022, Leah Crane, Odd waves within sun defy explanation

clxxx Planetary Society podcast, planetary.org/multimedia/planetary-radio/show/, We have touched the Sun: The Parker Solar Probe's triumph, published 12 January 2022

clxxxi Astronomy April 2021, The Anatomy of the Sun

clxxxii Astronomy February 2021, Alison Klesman & Jake Parks, Top 10 Space Stories of 2020 - Solar Science Enters a Golden Age

clxxxiii Astronomy April 2022, Caitlyn Buongiorno, Quick Takes – Defining a Planet

clxxxiv Astronomy August 2022, Mark Zastrow, Protoplanet shakes up formation theory

clxxxv New Scientist 2 October 2021, Jonathan O'Callaghan, Hints of a solar system smash-up

clxxxvi Astronomy May 2020, Alison Klesman, How the Planets Line Up

clxxxvii Astronomy February 2021, Alison Klesman, Planetary Surface Pressures

clxxxviii Astronomy October 2021, Stephen James O'Meara, Catching Mercury by the tail

clxxxix How the Universe Works, Series 9 episode 3, Secrets of the Sun, aired 7 April 2021

cxc Astronomy December 2021, Caitlyn Buongiorno, Ask Astro

cxci New Scientist 21 August 2021, Leah Crane, Mercury has almost no boulders and we're not sure why

cxcii New Scientist 8 May 2021, Jonathan O'Callaghan, Venus Finally

Reveals Some Inner secrets

[cxciii] How the Universe Works, Season 4 episode 9, The Planet of Nightmares, aired 19 May 2017

[cxciv] How the Universe Works, Season 2 episode 2, Megastorms – The Winds of Creation, aired 18 July 2012

[cxcv] New Scientist 23 October 2021, Jonathan O'Callaghan, Venus's surface may always have been too hot for oceans

[cxcvi] Astronomy December 2021, William Sheehan, Unveiling the Clouds of Venus

[cxcvii] The Universe, Season 4 episode 4, Biggest Blasts, aired 8 September 2009

[cxcviii] How the Universe Works, Season 5 episode 3, Black Holes: The Secret Origin, aired 6 December 2016

[cxcix] New Scientist 9 October 2021, Deep gorges on Mars sign of potent floods

[cc] Astronomy December 2021, Jake Parks, Cosmic tour of the planets

[cci] Astronomy February 2021, Alison Klesman & Jake Parks, Top 10 Space Stories of 2020 - Mission to Mars

[ccii] Astronomy July 2021, Hot Bytes – Gone to Ground

[cciii] New Scientist 12 February 2022, Chen Ly, Mars pummelled by asteroids for longer

[cciv] Astronomy February 2021, Alison Klesman and Jake Parks, Mars and Venus are Geologically Active

[ccv] New Scientist 25 September 2021, Leah Crane, Mushballs in the atmospheres of the ice giants

[ccvi] New Scientist 5 February 2022, Will Gater, An explanation for the different blues of Uranus and Neptune

[ccvii] New Scientist 17 July 2021, Leah Crane, Mystery of giant planet's rays solved

[ccviii] Astronomy December 2021-12, Mark Zastrow, Jupiter's Aurorae Trigger Heat Waves

[ccix] New Scientist 22 May 2021, Jonathan O'Callaghan, Europa's secret lakes may host life

[ccx] Astronomy March 2021, Caitlyn Buongiorno, Jupiter's Luminescent Moon

[ccxi] Astronomy December 2021, Jake Parks, Quick Takes – Saturn's Hazy Heart

ccxii New Scientist 22 January 2022, Leah Crane, Saturn's small moon Mimas may have an impossible ocean
ccxiii New Scientist 30 October 2021, Leah Crane, Titan may be doomed to fly away or smash into Saturn
ccxiv New Scientist 20 November 2021, Leah Crane, Biggest moons of Uranus may have subsurface oceans
ccxv New Scientist 26 June 2021, Leah Crane, Mystery of the red patches on Pluto
ccxvi New Scientist 9 April 2022, Leah Crane, Pluto may have active ice volcanoes
ccxvii Astronomy June 2021, William Jennings, Ask Astro
ccxviii New Scientist 20 November 2021, Jonathan O'Callaghan, Enigmatic Planet Nine may have been seen by a space telescope in 1980s
ccxix New Scientist 28 August 2021, Leah Crane, Half a million new asteroids found in asteroid belt
ccxx Astronomy May 2020, Richard Talcott, A Thousand Points of Light
ccxxi The Universe, Series 2 episode 12, Cosmic Collisions, aired 4 March 2008
ccxxii Wikipedia, Baptistina Family, accessed 9 January 2022
ccxxiii New Scientist 23 October 2021, Leah Crane, The Lucy spacecraft has set off on a journey to study Trojan asteroids
ccxxiv Planetary Society podcast, planetary.org/multimedia/planetary-radio/show/, A good year for space: Planetary Society all-stars review 2021, published 29 December 2021
ccxxv New Scientist 20 November 2021, Leah Crane, Moon chunk may be Earth's space buddy
ccxxvi Wikipedia, Earth trojan, accessed 12 December 2021
ccxxvii New Scientist 27 November 2021, Leah Crane, A second trojan asteroid share's Earth's orbit
ccxxviii Astronomy May 2020, Leslie Nemo & Jake Parks, Found: Crater from 790,000-Year-Old Asteroid strike
ccxxix The Universe, Series 4 episode 3, It Fell from Space, aired 1 September 2009
ccxxx Astronomy July 2021, Jake Parks, Quick Takes – Apophis Will Pass

ccxxxi Astronomy March 2021, S. Alan Stern, Mission to the Centaurs

ccxxxii New Scientist 16 October 2021, Leah Crane, NASA spots odd pair of asteroids in deep space

ccxxxiii Astronomy June 2021, Hot Bytes – Aptly Named

ccxxxiv New Scientist 12 February 2022, Jonathan O'Callaghan, 'Mega comet' from outer solar system is 137 kilometers wide

ccxxxv How the Universe Works, Series 2 episode 6, Comets – Frozen Wanderers, aired 15 August 2012

ccxxxvi Astronomy July 2021, Caitlyn Buongiorno, 'Oumuamua's Trip Around the Sun

ccxxxvii Astronomy February 2021, Alison Klesman & Jake Parks, 10 Top Space Stories from 2020, A Triple Comet Surprise

ccxxxviii Astronomy June 2021, Alison Klesman, Earth is a Planet Too!

ccxxxix The Great Courses, Prof. Alex Filippenko, Understanding the Universe: An Introduction to Astronomy, 2nd Edition, published 2007

ccxl New Scientist 7 May 2022, Robin George Andrews, A seismic mystery

ccxli Astronomy March 2021, Mark Zastrow, Making the Moon

ccxlii Astronomy March 2021, William K. Hartmann, Ask Astro

ccxliii New Scientist 19 March 2022, Alex Wilkins, Moon's emergence from planet crash reconstructed

ccxliv New Scientist 19 March 2022, Jonathan O'Callaghan, Double-shadowed craters could hold ice on the moon

ccxlv New Scientist 16 October 2021, Leah Crane, Moon spewed lava relatively recently

ccxlvi New Scientist 12 December 2020, Jonathan O'Callaghan, Hot rocks may have given Mars its water

ccxlvii New Scientist 14 August 2021, Matthew Sparkes, Moon may have always had a puny magnetic field

ccxlviii Astronomy April 2020, John Wenz, Visit the Nearest 14 Habitable Exoplanets

ccxlix Astronomy July 2021, Jake Parks, Quick Takes – New Vega

ccl Astronomy April 2020, Jake Parks, Giant Planet Found Around White Dwarf

ccli Astronomy May 2021, Mark Zastrow & Jake Parks, Quick Takes

cclii New Scientist 22 January 2022, Alex Wilkins, Super-Earths have life-friendly shields

[ccliii] Astronomy June 2021, Alison Klesman, A Nearby Exoplanet with an Earth-Like Atmosphere

[ccliv] Astronomy July 2021, Arwen Rimmer, Volcanoes may have replenished a super-Earth's atmosphere

[cclv] Wikipedia, Proxima Centauri b, accessed 27 December 2021

[cclvi] Astronomy June 2022, Alison Klesman, Third planet found around Proxima Centauri

[cclvii] Wikipedia, TRAPPIST-1, accessed 27 December 2021

[cclviii] New Scientist 31 July 2021, Leah Crane, Birth of an alien moon glimpsed for first time

[cclix] New Scientist 23 October 2021, Leah Crane, Odd space signal may be shape shifting cloud

[cclx] Astronomy March 2021, Dominique Petit de la Roche, Ask Astro

[cclxi] Astronomy April 2021, Caitlyn Buongiorno, Far-Flung Exoplanet Resembles Long-Sought Planet Nine

[cclxii] Astronomy February 2021, Caitlyn Buongiorno, Planet Captured in the Light of a Dying Star

[cclxiii] New Scientist 23 October 2021, Chen Ly, Giant exoplanet shrugs off death star

[cclxiv] New Scientist 19 February 2022, Leah Crane, Habitable world may be orbiting dead star

[cclxv] Astronomy July 2021, Jake Parks, Quick Takes – Two-Faced

[cclxvi] Astronomy April 2020, Jake Parks, Quick Takes - Super Puffs

[cclxvii] New Scientist 18 December 2021, Chen Ly, 'Impossible' planet orbits binary stars

[cclxviii] Astronomy February 2021, Caitlyn Buongiorno, Twisted Solar System

[cclxix] Astronomy February 2022, Alison Klesman, An Exoplanet Orbiting Three Stars?

[cclxx] Wikipedia, Tabby's Star, accessed 27 December 2021

[cclxxi] New Scientist 22 January 2022, Leah Crane, Astronomers may have found a huge moon around a Jupiter-like exoplanet

[cclxxii] Astronomy April 2021, Randall Hyman, The Galaxy's Marvelous Rogues and Misfits

[cclxxiii] New Scientist 25 September 2021, Leah Crane, Potential exomoon found orbiting a giant rogue planet cclxxiv New Scientist 19 June 2021, Building blocks for life at galaxy's edge

cclxxv New Scientist 25 September 2021, Leah Crane, Maps of planet-forming zones will help the hunt for alien life

cclxxvi New Scientist 30 April 2022, Carissa Wong, DNA bases found in space rocks

cclxxvii Astronomy January 2021, David A. Kring, The Origins of Life on Earth

cclxxviii New Scientist 19 June 2021, Katherine Sanderson, A robotic chemist could reveal the recipe for Earth's primordial soup

cclxxix New Scientist 14 August 2021, Danielle Sedbrook, Organic blobs built in lab may be a step towards synthetic life

cclxxx New Scientist 14 July 2009, Timeline: The Evolution of Life, Michael Marshall

cclxxxi Wikipedia, Evolution, accessed August 2021

cclxxxii The Great Courses, Prof. Laird Close, Life in Our Universe, published 2013

cclxxxiii New Scientist 7 August 2021, Michael Le Page, Single-celled organism has evolved a natural mechanical computer

cclxxxiv New Scientist 26 September, Carrie Arnold, Evolution Evolving – Genes Aren't Destiny

cclxxxv New Scientist 21 August 2021, Michael Marshall, Stripped-back cell still evolves

cclxxxvi Nova, Series 47 episode 10, The Secret Mind of Slime, aired 16 September 2020

cclxxxvii New Scientist 16 October 2021, Michael Marshall, Lava-munching microbes were among the earliest things on land

cclxxxviii New Scientist 24 July 2021, Claire Ainsworth, 'Borg' DNA assimilates genes

cclxxxix New Scientist 7 August 2021, Michael Marshall, Oldest ever animal fossils discovered

ccxc New Scientist 15 January 2022, Jason Arunn Murugesu, Mystery of oxygen made by sea microbe

ccxci New Scientist 8 May 2021, Michael Marshall, Billion-year-old microbe took steps towards internal organs

ccxcii New Scientist 19 June 2021, Riley Black, Early land dwellers returned to water [ccxciii] New Scientist 16 October 2021, Michael Benton, China's dinos

[ccxciv] New Scientist 11 December 2021, Michael Marshall, Fossil footprints hint at mystery hominin with unusual walking style

[ccxcv] New Scientist 27 November 2021, Michael Marshall, Hominin had curved spine that helped with upright walking

[ccxcvi] Wikipedia, Human Evolution, accessed 14 February 2022

[ccxcvii] Nova, Series 6 episode 1, What Makes Us Human?, aired 10 October 2012

[ccxcviii] Origins of Humankind, Series 1 episode 1, We Are Fire Creatures from an Ice Age, aired 12 March 2017

[ccxcix] New Scientist 5 February 2022, Michael Marshall, Fossil skull may be Denisovan

[ccc] New Scientist 13 November 2021, Michael Marshall, Infant skull suggests Homo naledi buried their dead 250,000 years ago

[ccci] Doom, by Niall Furguson, published 2021 by Penguin Press

[cccii] New Scientist 27 November 2021, Kate Ravilious, The last human

[ccciii] Wikipedia, List of Extinction Events, accessed 14 February 2022

[ccciv] How the Universe Works, Series 3 episode 6, Weapons of Mass Extinction, aired 13 August 2014

[cccv] New Scientist 27 November 2021, Really Brief – Extinction began with volcanic winter

[cccvi] New Scientist 23 October 2021, Bas den Hond, Asteroid impact in vivid details

[cccvii] New Scientist 29 January 2022, Adam Vaughan, Hunting the Anthropocene's dawn

[cccviii] Astronomy Magazine Special October 2020, Michael R. Rampino, Cosmos – Origin and Fate of the Universe, Dark Matter's Shadowy Effect on Earth

[cccix] Discover, November 2020, Eric Betz, The Wandering Stars That Pass by Our Solar System

[cccx] UniverseToday.com, A Supernova 2.6 Million Years Ago Could Have Wiped Out the Ocean's Large Animals

[cccxi] Astronomy August 2021, Alison Klesman, Oxygen may not be a sure sign of life

[cccxii] Astronomy January 2021, Morgan L. Cable, Looking for Life in

the Universe

cccxiii Astronomy April 2020, Bob Berman, In Praise of Nothing

cccxiv New Scientist 22 January 2022, Leah Crane, Organics in martian rock not made by life

cccxv New Scientist 24 July 2021, Jonathan O'Callaghan, Methane-burping microbes may live near Curiosity rover

cccxvi New Scientist 4 September 2021, Leah Crane, Martian cave entrances could be friendly for life

cccxvii New Scientist 2 October 2021, Jason Arunn Murugesu, Window for life on Venus narrows to first billion years after it formed

cccxviii New Scientist 3 July 2021, Leah Crane, Clouds on Venus don't have enough water to support life

cccxix Astronomy February 2021, Alison Klesman & Jake Parks, Top 10 Space Stories of 2020, Astronomers Spy Phosphine on Venus

cccxx Astronomy May 2021, Mark Zastrow & Jake Parks, Quick Takes

cccxxi Astronomy February 2021, Mark Zastrow, Phosphine on Venus?

cccxxii New Scientist 17 July 2021, Leah Crane, Mystery Venus gas may be volcanic

cccxxiii Are We Alone, published 2021 by Bauer Media Group

cccxxiv New Scientist 20 November 2021, Abigail Beall, Tell Me Why – Why haven't we heard from aliens?

cccxxv How the Universe Works, Series 3 episode 5, Is Saturn Alive?, aired 6 August 2014

cccxxvi Through the Wormhole, Series 2 episode 10, What Do Aliens Look Like?, aired 3 August 2011

cccxxvii New Scientist 20 November 2020, Richard Webb, Tell Me Why – Why is the universe just right?

cccxxviii New Scientist 11 September 2021, Thomas Lewton, Stephon Alexander Interview: Is the universe a self-learning AI?

cccxxix New Scientist 6 October 2018, Michael Brooks, The wrong number

cccxxx New Scientist 2 May 2020, Michael Brooks, "Here. There. Everywhere?"

cccxxxi New Scientist 2 October 2021, Alan Wells, Letters – Maybe our planet is a neural network too

cccxxxii Through the Wormhole, Series 5 episode 5, Does the Ocean

Think?, 18 June 2014

[cccxxxiii] Astronomy March 2022, Caitlyn Buongiorno, Ask Astro – Cosmic Fireworks

[cccxxxiv] Astronomy April 2021, Matt Caplan, Earth's Ultimate Fate

[cccxxxv] Astronomy January 2021, Eric Betz, The Big Crunch vs. The Big Freeze

[cccxxxvi] The End of Everything, Katie Mack, published 2020 by Scribner (Simon and Schuster)

[cccxxxvii] New Scientist 4 December 2021, Leah Crane, Mini black holes could spell trouble

[cccxxxviii] New Scientist 18 August 2018, Chelsea Whyte, A glimpse of a previous universe

[cccxxxix] New Scientist 8 May 2021, Matthew Sparkles, Frigid molecules act as a single quantum object

[cccxl] Wikipedia, Superconductor, accessed 14 February 2022

[cccxli] Wikipedia, Superconductivity, accessed 14 February 2022

[cccxlii] Robert Ferguson Observatory, Santa Rosa, CA (rfo.org)

[cccxliii] The Great Courses, Prof. Frank Summers, New Frontiers: Modern Perspectives on the Solar System, published 2008

[cccxliv] The Great Courses, Prof. Bradley E Schaefer, The Remarkable Science of Ancient Astronomy, published 2017

[cccxlv] Astronomy December 2021, Caitlyn Buongiorno, Sodium May Make Asteroid Phaethon Fizz

[cccxlvi] Astronomy July 2021, Alison Klesman, Mars may generate the zodiacal light

www.ingramcontent.com/pod-product-compliance
Lightning Source LLC
Chambersburg PA
CBHW071301140726
47996CB00005B/1577